WATER POWER
AND WATERMILLS

WATER POWER AND WATERMILLS

An Historical Guide

JONATHAN BROWN

THE CROWOOD PRESS

First published in 2011 by
The Crowood Press Ltd
Ramsbury, Marlborough
Wiltshire SN8 2HR

www.crowood.com

© Jonathan Brown 2011

All rights reserved. No part of this publication may be reproduced or transmitted in any form or by any means, electronic or mechanical, including photocopy, recording, or any information storage and retrieval system, without permission in writing from the publishers.

British Library Cataloguing-in-Publication Data
A catalogue record for this book is available from the British Library.

ISBN 978 1 84797 243 9

Frontispiece: Horstead Mill, Norfolk, an example of the big commercial mills of eastern England, with weather-boarded external cladding. Photographed about 1960, the mill burned down in 1963. (Museum of English Rural Life)

Typeset by Phoenix Typesetting, Auldgirth, Dumfriesshire.

Printed and bound in Singapore by Craft Print International Ltd.

Contents

	List of Illustrations	6
	Preface	10
1	Introduction	11
2	Where Did the Mills Come From?	15
3	The Medieval Mill	20
4	The Technology of Water Power	25
5	Flour-Milling Before the Industrial Revolution	54
6	Early Industrial Uses of Water Power	57
7	Water Power and the Industrial Revolution	73
8	Developing the Technology of Water Power	88
9	Flour-Milling in the Industrial Economy	127
10	Water Power on the Farm	135
11	Water Power and the Competition from Steam	138
12	Landscapes of Water Power	161
13	Hydro-Electricity	169
14	A Green Future for Water Power?	177
15	Conclusion	186
	Notes	187
	Glossary	197
	Bibliography	199
	Index	205

List of Illustrations

Frontispiece: Horstead Mill, Norfolk.
p.11: The weir at Belper, Derbyshire.
p.12: A hand quern of British origin.
p.13: A small hand mill for farm use.
p.14: Flatford Mill on the River Stour in Suffolk.
p.15: The Archimedean screw.
p.16: Diagram of a noria.
p.17: The basic layout of a watermill as described by Vitruvius.
p.21: The mill at Houghton in Huntingdonshire.
p.24: The medieval mill building at Fountains Abbey.
p.26: The operation of a horizontal mill wheel.
p.27: A horizontal wheel illustrated in 1851.
p.28: The different types of vertical waterwheel.
p.29: An undershot water wheel at Mapledurham Mill.
p.29: Twin overshot wheels at Dunster Mill, Somerset.
p.30: Angled and elbow buckets.
p.31: The construction of a clasp-arm wheel.
p.31: Two large diameter wheels at Gweek, near Helston, Cornwall.
p.33: Eversley New Mill, Hampshire.
p.34: The siting of mills in the mainstream of the river.
p.35: A weir at New Mills, Cheshire.
p.35: The weir across the River Derwent at Chatsworth.
p.35: Weir and sluice gate at New Mills.
p.36: The weir facing downstream at Masson Mills, Derbyshire.
p.36: Arrangement of water engineering for a mill on a leat off the river.
p.37: The weir and sluice gate at New Mills.

p.38: A launder carrying water to an overshot wheel at Synod Mill, Cardiganshire.
p.39: The millpond for the corn mill at Cromford, Derbyshire.
p.39: Old wooden sluice gates, Belvedere Mill, Chalford, Gloucestershire.
p.40: Inflow gate and debris filter, Lea Bridge, Derbyshire.
p.40: Overflow channel at Tregwynt woollen mill, Pembrokeshire, 1959.
p.41: The weir at Ruswarp, north Yorkshire.
p.42: Weir at Quarry Bank Mill, Cheshire.
p.45: Drive from cogwheel to lantern pinion, or trundle wheel.
p.46: Hand mill with lantern drive.
p.46: Nether Alderley Mill, Cheshire.
p.47: Diagram showing the working of a cam wheel to operate fulling stocks.
p.48: The watermill in the Luttrell psalter.
p.49: The medieval mill at Abbotsbury, Dorset, known now as the Old Malthouse.
p.50: Rossett Mill, near Wrexham.
p.51: Thorrington tide mill, Essex.
p.51: The undershot wheel at Thorrington.
p.52: Map of Thorrington and St Osyth tide mills in Essex.
p.52: The Three Mills, Bromley-by-Bow, east of London.
p.53: Eling tide mill on a creek off Southampton Water.
p.54: Ham Mill, Newbury, Berkshire.
p.55: Hele mill, near Ilfracombe, Devon.
p.56: Shetland Islands mill, 1940s.
p.59: Fulling stocks at Hunt & Winterbotham's mill, Cam, Gloucestershire.
p.60: Fulling mill interior, Helmshore, Lancashire.

List of Illustrations

p.61: Late medieval mill at West Harnham, near Salisbury.
p.62: Loudwater Mill, Buckinghamshire.
p.63: Washford silk mill, Congleton.
p.64: The silk mill at Whitchurch, Hampshire.
p.66: A water-powered trip hammer at Holbeam, Devon.
p.67: The needle mill at Redditch, Worcestershire.
p.69: Bere Mill, Laverstoke, Hampshire.
p.70: Water-powered edge-runner stones for mixing gunpowder.
p.72: Sorocold's water wheel for the London Bridge waterworks from *The Universal Magazine* of 1749.
p.75: Arkwright's water spinning frame, as illustrated in Rees's *Encyclopaedia*.
p.76: Torr Vale Mill, New Mills in Cheshire, built 1788.
p.77: Quarry Bank Mill, Styal.
p.79: Belvedere Mill, Chalford, Gloucestershire.
p.80: Inside the spinning shed of a Welsh woollen mill powered by water in the 1930s.
p.83: Overshot wheel at Finch Foundry, 1966.
p.84: Forge hammers at Finch Foundry, Sticklepath, Devon, 1966.
p.85: Sarehole Mill, south of Birmingham.
p.86: A sawmill built for Chatham dockyard early in the nineteenth century.
p.89: A breast shot wheel designed by John Smeaton, illustrated in Rees's *Cyclopaedia*.
p.90: A high-breast wheel, constructed of iron by Sir William Fairbairn for a mill at Cleator, near Whitehaven.
p.91: Curved floats fitted to the water wheel at Terwick mill, Rogate, Sussex.
p.91: A pitchback wheel, illustrated by Abraham Rees.
p.92: The common form of lifting sluice gate, which allows the water to flow underneath.
p.92: Improved sluice gates for waterwheels.
p.93: Fairbairn's improved ventilated buckets.
p.94: The Poncelet waterwheel, Buscot Park, Berkshire, 1866.
p.95: Waterwheel built using light wrought iron spokes.
p.96: A drawing of a tin mine at Botallack, Cornwall, showing one of the large water wheels used for draining and working ore.
p.96: The great wheel at Laxey, Isle of Man.
p.97: The North Mill at Belper from Rees's *Cyclopaedia*.
p.98: Wide breast-shot wheel, Lower Slaughter, Gloucestershire.
p.98: Water wheel built for Waltham Abbey ordnance factory.
p.99: Wheel house, Lumbutts Mill, Todmorden, Yorkshire.
p.100: The construction of an underground tail race.
p.101: The wooden breast-shot wheel in Longbridge corn mill, Hampshire.
p.102: The construction of a cast-iron axle for a waterwheel.
p.103: This iron wheel was used at the corn mill, Cromford, Derbyshire.
p.103: An iron pitch-back wheel at Castlebridge Mill, Wexford, in 1938.
p.104: Hybrid wood and iron wheel, Pillmire Bridge, near Reeth, Yorkshire.
p.104: Suspension wheel built for the cotton mills at Catrine.
p.105: Improved low-breast waterwheel built in the late 1820s.
p.106: New wheels for old at Cobham Mill, Surrey, 1953.
p.107: The corn mill at Nuneaton with two pairs of stones driven from the one waterwheel, illustrated by Henry Beighton in 1723.
p.108: Plan of the layout of drive for a horizontal shaft, as depicted by Beighton.
p.109: The layout of a typical layshaft drive running in line with the axle of the water wheel, from Joseph Glynn's treatise on water power.
p.109: Layshaft drive in Hockley Mill, near Winchester, Hampshire.
p.110: The basic arrangement of drive in a corn mill using an upright shaft and spur wheel.
p.110: Gearing in Sarehole Mill, near Birmingham, a mixture of iron and wood construction.
p.111: The pitwheel and drive in Marcham Mill, near Banbury, 1953.

List of Illustrations

p.112: A bolting machine for sifting and grading flour.

p.113: A bevelled gear wheel made of wood in Eling tide mill.

p.115: Dane in Shaw Mill, outside Congleton, built in 1784.

p.116: Dr Barker's wheel, illustrated by Joseph Glynn.

p.117: Outline diagram of a Fourneyron turbine.

p.118: The Vortex turbine illustrated in Williamson Brothers' catalogues of the 1870s.

p.119: The installation of a Vortex turbine in the millstream and drive to the machinery.

p.120: Francis turbine manufactured by Armfield of Ringwood.

p.120: The flow of water through a turbine.

p.121: Pelton wheel made by Charles Hett.

p.121: A Pelton wheel at a tinplate rolling mill at Cwm Avon near Port Talbot.

p.122: A hydraulic engine for winding at a lead mine near Richmond, Yorkshire.

p.122: Hydram hydraulic ram from catalogue by John Blake Ltd, 1950.

p.123: The position of mill wheels.

p.124: The fulling mill at Helmshore, Lancashire.

p.125: The Masson Mills outside Cromford, built for Sir Richard Arkwright.

p.125: Venetian windows, Stanley Mill, Gloucestershire.

p.126: The flour mill at Chatsworth.

p.126: Piccotts End Mill, Hertfordshire.

p.128: Sturminster Newton Mill, Dorset.

p.129: Daniel's Mill, near Bridgnorth, Shropshire.

p.130: Whitmore & Binyon catalogue for water wheels, 1876.

p.131: Four pairs of stones at Longbridge Mill, Hampshire.

p.132: One of Norfolk's grand mills, at Buxton.

p.133: Tile Mill on the River Kennet in Berkshire in the 1950s.

p.134: Burghfield Mill on the River Kennet in Berkshire.

p.135: The mill at Coley Farm, Gnossal, Staffordshire, built in the 1840s.

p.136: A wooden water wheel for pumping on Nethercote Farm, Bourton-on-the-Water, Gloucestershire, in use in 1943.

p.141: Savery steam pump to pump water from the stream back up to the waterwheel.

p.147: Gibson Mill, Hebden Bridge, Yorkshire.

p.148: Tending the spinning machines in Richmond Mill, Huntly, Aberdeenshire, still water-powered in the mid-twentieth century.

p.149: Upper Greenland Mill, the last working woollen mill in Bradford-upon-Avon.

p.150: A Welsh woollen mill, 1937.

p.151: Witchampton paper mill in Dorset. Photograph by Alan Stoyel.

p.152: Churchill Forge, Worcestershire.

p.153: A water-powered ore crusher, an illustration from Tomlinson's *Cyclopaedia*.

p.154: The wheel built for the Weardale Lead Co., at Kilhope, Co. Durham.

p.156: Caudwell's Mill, Rowsley, Derbyshire, one of those that installed turbine-powered roller mills.

p.157: A. R. Tattersall & Co., of London, catalogue for the 'Midget' roller mill, 1915.

p.158: Bryncrug corn mill, Merionethshire, built in the sixteenth century, was still in use in 1938, run by Miss Lois Owen.

p.158: Mr Lightfoot, the last miller at Hessenford, Cornwall, in 1957, when he retired aged 75.

p.159: The watermill at Aymestrey, Herefordshire, being restored for use during the Second World War.

p.161: The country mill: Ramsbury, Wiltshire in the 1930s.

p.162: The Holy Brook flows through Calcot on its way towards Reading.

p.163: The course of the Holy Brook from Theale to Reading.

p.164: Map of the River Itchen and its branches and the siting of mills around Winchester.

p.165: A row of cottages built near Quarry Bank Mill to house workers.

p.166: Map of the impact of water power on the river landscape at Belper.

p.168: Modern environmental landscaping at Glendoe reservoir.

p.169: Overshot water wheel used for an electricity

List of Illustrations

- generation scheme started in 1914 by the Glanllyn estate in Merionethshire.
- p.173: Pitlochry dam and hydro-electric power station on the River Tummel, seen in 1954.
- p.174: Fasnakyle dam, part of the Glen Affric power development.
- p.175: Tummel power station.
- p.177: A mill in decay at Sawbridgeworth, Hertfordshire, in 1980.
- p.178: Newnham Mill, Northamptonshire, being demolished in August 1954.
- p.178: Ixworth Mill, Suffolk, in a sorry-looking state in the mid-twentieth century.
- p.179: Woodbridge tide mill in 1957, since preserved in working order.
- p.179: Overshot water wheel at the preserved Finch foundry.
- p.179: Mill converted to a house at Grindleford, Derbyshire.
- p.180: Mapledurham Mill on the River Thames.
- p.181: The reservoir at Glendoe in the midst of a Highland winter.
- p.183: Torrs Hydro's micro-power generator at New Mills.
- p.184: Aquamarine Power's Oyster hydro-electric wave energy converter.
- p.185: Diagram of the Oyster wave energy converter in operation.
- p.186: The waterwheel at Coley Farm, Staffordshire.

Preface

Authors often mention their publisher somewhere towards the end of the acknowledgements, but I would like to start by thanking The Crowood Press for inviting me to do this book. I did not jump at the chance. I acted responsibly – I went away and thought about it, and *then* jumped at it. I was daunted a bit by the prospect of fitting such a wide subject into the fairly narrow compass of this book, but it is a wonderful story, and one in which I have had an interest for many years. The mills of the village and the country town, and of the more urban landscape, have been a part of my work at the Museum of English Rural Life almost from my first day there (which was some time ago now). In my student days, a project on Scottish hydro-electric power took me on a hiking tour with my brother to visit many of the sites in the Highlands. At the end the feet were sore and the back ached, but the experience had been great. The following year, we went on our bicycles.

This book represents something of a drawing together of my studies and of my interests that have developed over the years, but it remains nevertheless more a synthesis of work published elsewhere than a detailed piece of original documentary research. To all those who have written books, journal articles and web content relating to the subject, I owe a great debt of gratitude. Of necessity, much has to be left out of a book of this scope, and it is to all those others that the reader should turn for additional detail.

One of the pleasures of writing this book has been the opportunity to visit working mills, most of them open as museums. The enthusiasm of the staff and volunteers whom I have met has made those visits a joy, and my thanks go to all of them. One of them was Gavin Bowie, an old friend, with whom a conversation in a freezing cold Hampshire mill one New Year's Day proved very stimulating. The members of the Mills Section of the Society for the Protection of Ancient Buildings, and also Ron Cookson, Luke Bonwick and the others at the Mills Archive Trust, have been very welcoming and helpful. The Esk Valley Community Energy Group in Yorkshire made me most welcome when I went as a visitor to their public meeting, generously explaining their plans for a hydro-electric generator on the river. Sara Parsons of the National Trust, Hardcastle Crags, was very helpful in her answers to my questions, as were the public relations departments of power companies, including Scottish & Southern, First Hydro and Aquamarine Power.

My colleagues and friends at the Museum of English Rural Life have, as ever, been a source of strength and encouragement, with Caroline Benson again worth her weight in gold on the photographic side of things. My brother read the manuscript and guided me through many technical matters and, last, but certainly not least, my wife Patricia not only read the manuscript but also prepared most of the drawings.

1 Introduction

To see the great weir at Belper in Derbyshire is to gain some idea of the importance that water power has played in economic development. Even though it is now divorced from the waterwheels and mills that it was built to serve, it remains an impressive sight. And it was part of a much bigger complex of channels constructed to serve mills, most of which have disappeared. Sites such as these emphasize how much of the world's economic, technical and social development has been bound up in energy, from muscle power to atomic power. Even casual reading of the current debate reveals the importance accorded to energy – its sources, its security and its cost. For centuries, one of the prime sources of energy has been water, with the downward flow of river water and the waves on the tide being an obvious potential provider of power.

For some historians of civilization, the watermill represents one of the prime steps forward, especially for those whose approach to the development of

The weir at Belper, Derbyshire, built to divert water from the River Derwent to the cotton mills.

Introduction

A hand quern of British origin. (Museum of English Rural Life)

technology is broadly evolutionary. Lewis Mumford, in his study published in 1934, considered the transition from manual to water power and horse power to be one of the main characteristics of his 'eotechnic' phase of development. That was superseded by the 'paleotechnic' phase (coal and steam) and the 'neotechnic' (electricity). R. J. Forbes divided technological advance into five phases, dominated respectively by the power of: human muscle; human and animal power; water power; steam power; and the nuclear age, which was just beginning as he was writing, in 1955.[1]

Water was the first inanimate power source to be harnessed and the introduction of water power represented a considerable gain in energy. Grinding grain by hand is extremely laborious: using the oldest type of hand quern, the saddle quern, it can take two hours to grind about half a pound of flour. The rotary quern, introduced in the fifth century BC and reaching Britain about 400BC, was more productive: two women working it could grind about 10 pounds of flour in an hour. Later, geared mills of various types were developed, and a smaller, hand-powered version of the early nineteenth century could grind 40 to 50lb. Such mills were common on British farms, mainly for preparing animal feed, and in many parts of the world they are still used for grinding meal and flour.

The use of animals provided significantly more power – a horse is about eight times more powerful than a man, an ox about five times – and, as heavier grindstones were introduced into the ancient world, animal-powered mills became more common. The waterwheel could provide even more power. A medieval watermill of average size had a grinding capacity of about five times that of a two-man hand mill.

So, the use of water power could save labour, it could increase production, and it could enable work to be done that otherwise might not be possible, such as pumping water out of deep mines. But such benefits are only worth having if there is a pay-back on the investment in the technology, and one of the features of water power that recurs throughout its history is its high capital cost. Not only is the wheel, the prime mover, costly, but also the associated waterworks and gearing. The technologies to harness the power of water also needed to be sufficiently developed before they could be applied. These issues largely explain the uneven rate at which water power was taken up. There are many examples of industries that were apparently well suited to using water power failing to do so until long after it was a feasibility; in other cases, water power might be taken up in one part of the country, but not elsewhere.

The principle behind water power is a simple one: water flowing downhill turns a wheel, which in turn drives a machine. The smallest corn mill and the largest hydro-electric power station share that basic driving force. Some of the wheels are small, some enormous, some turn in the horizontal plane, others are vertical, but, turned by water, each produces usable power in the same way. Rivers flow downwards to the sea and the mill engineer exploits that to create a 'head' or 'fall' sufficient to turn a wheel. This usually involves diverting water from the stream into a millrace to feed the wheel; a millpond is often built to store water. Wherever water power is tapped a head of water has to be created – sometimes it is only a few feet, at other sites a head of tens of feet can be built up, while a head of hundreds of feet is employed at some power stations. The most prized sites, therefore, were those for which the least amount of engineering was needed to build up the head.

For the first dozen centuries or so, water power was primarily employed for the milling of flour. This was the use to which the Romans and other peoples first applied the technology, and this continued

A small mill for farm use, designed to be driven by hand. This mill was made in the late eighteenth century and used on a farm in Essex. (Museum of English Rural Life)

throughout the medieval period, even after water power had become established more generally in industry and mining. In the production of flour it was the mill that ground the grain, and the waterwheel that drove it. It was a natural step for the whole of the works, from wheel to millstones, and the building that housed them, to be referred to as a 'watermill'. The significance of flour-milling in the development of the use of water power was such that its terminology came to be adapted to textile manufacture, paper-making and other works, as new applications for the power were found. The word 'watermill' gained a number of variations of meaning, while the term 'mill', which originally meant the corn-grinding mechanism, came to be applied to the building that housed the wheel and millstones and thence to factories, especially those producing textiles, regardless of their source of power.

As manufacturing industry and mining grew, so did their need for power, and they found it, as often as not, in the waterwheel. Without water power, many significant developments in technology would have been delayed or limited, or perhaps even stillborn. Water power had a major role to play in the surge of economic activity in the Middle Ages, and the early stages of industrialization in the eighteenth century relied heavily on water power. Steam engines took some time to become established; indeed, some historians have argued that there was a 'crisis of power' in the early nineteenth century as a result of the slow pace of development of steam, and that water power proved some salvation.[2] Once steam engines did prove their value, they and other forms of heat engine driven by gas or oil seemed to eclipse the power of water in the nineteenth and twentieth centuries. Water power was transformed again, however, with the development of the water turbine and the introduction of hydro-electric power. A considerable proportion of the world's electricity is now generated by the power of water. Nowadays water power is often classed as 'alternative technology', a sustainable and renewable resource; as such it might be reckoned to have a great future.

Water power has proved extremely adaptable. Indeed, this flexibility is imprinted into many an old mill site, the uses of which have changed as the fortunes of different industries have fluctuated. A watermill might have gone from grinding corn to fulling cloth to making paper and back again, before finally retiring as a superior residence. The King's Mills at Weston-on-Trent have followed this pattern. This mill site, recorded in Domesday Book, had been converted from corn mill to fulling mill by the early sixteenth century. In 1680 it was a paper mill, expanding to operate on both sides of the river by 1718. Its final use, in the nineteenth century, was to grind gypsum.[3]

The watermill is a romantic thing, especially in what might be called its primary manifestation, the corn mill. The old weather-boarded or brick mill, with its 'unresting wheel sending out its diamond jets of water', has attracted artist, writer and poet.[4] Medieval artists highlighted the aesthetic attractions of the watermill as much as the later romantic painters. According to John Constable, whose father

Introduction

Flatford Mill on the River Stour in Suffolk, one of the scenes that made John Constable a painter.

owned several mills along the River Stour on the Suffolk–Essex border, 'The sound of water escaping from mill-dams, old rotten banks, shiny posts and brickwork – these scenes made me a painter – and I am grateful.'[5] Constable's landscapes did much to establish the popular view of the picturesque old mill, which grew along with a romantic interest in the countryside.

Sir Uvedale Price, in his early nineteenth-century *Essays in the Picturesque*, described the appeal that many artists, from J. M. W. Turner onwards, sought to capture: 'Such is the extreme intricacy of wheels and the woodwork; such the singular variety of forms and of lights and shadows, of mosses and weather stains from the constant moisture, of plants springing from the rough joints of the stones; such the assemblage of everything which most conduces to picturesqueness, that even without the addition of water an old mill has the greatest charm for a painter.'[6] The mill could also be a place of enchantment for children. Edmund Blunden was one writer who recalled being drawn as a child to the millstream and the fish in the pond, and simply sitting and watching the life around the mill at Cheveney, Kent, where the wheel was then generating electricity.[7]

The aura of romance was sometimes extended by contemporaries to the water-powered cotton mills of Arkwright or Strutt. On the whole, however, the new industrial mills were admired rather as marvels of the age, for the scale of their construction and the greatness of their power. When Joseph Wright of Derby painted Arkwright's mill at Cromford in 1789, it was the drama of its setting at night that attracted him. As that industrial age has receded, appreciation of surviving examples of the technology of the period has changed again, with a return, in part, of the romantic, and a new regard for the history of technological development.

2 Where Did the Mills Come From?

A HISTORY

Estimates of when men first started using water power have ranged back and forth with different attempts to interpret limited evidence. The early civilizations of the Middle East developed a number of systems for irrigation and water supply, including dams and reservoirs, canals, and devices for lifting water. The Hellenistic kingdom in north Africa, centred on Alexandria, was a particularly fertile area for inventiveness. This was where Archimedes lived, whose name remains associated with the water screw that was introduced in about 250BC. More recently, the Archimedean screw has been applied to the harnessing of power for the generation of electricity, but its original use was as a displacement pump for lifting water.

In common with most other water-lifting devices, the screw pump was driven by hand or animal power. There was one exception, the noria. This was a wheel with bucket compartments, which, when equipped with floats or paddles, could be turned by the force of the water – it was basically an undershot waterwheel. The buckets, made of bamboo, pottery or wood, scooped up water as they descended into the stream. The water was dropped into a trough as the buckets reached the top of the revolution (*see* overleaf)

Noria wheels could be large, up to 25 feet in diameter. Joseph Needham suggested that the noria originated in India in the fourth century BC, spreading thence to the Mediterranean and east to China. Not everyone agrees with that, but certainly it seems to have appeared in the West in the second or first century BC. Compared with the muscle-powered machines, the noria was probably uncommon at first, but its use certainly spread. After the Arab conquests it was known across north Africa and into Spain, and remained an important, if minor, use for water power.[8]

Whether there were any direct connections between the noria and the origins of the more familiar watermill has not been established. The chances are that they were developed independently, the tasks to which the two were put being quite different. The waterwheel applied to grinding corn depended on the prior development of the rotary mill, which had appeared in about the fifth century BC. This was widely used with animal-powered mills. The exact date when water power was applied to the rotary mill has remained a matter of conjecture. In about 25BC, the Roman architect Vitruvius described both the noria and a watermill driven by a vertical waterwheel in a book that has long been accepted as a prime source for the early watermill.

The Archimedean screw as designed for hand operation to lift water.

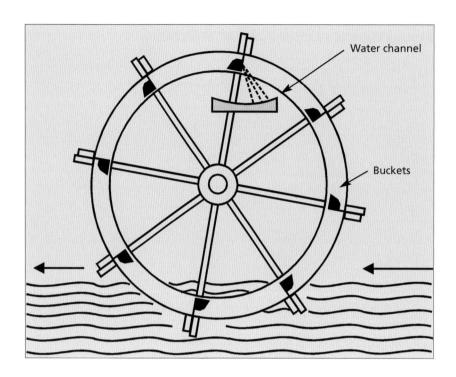

A simplified diagram of a noria, the water-powered water-lifting wheel. Pots on the side of the rim scooped up water to be tipped out at the top of the wheel's revolution into a cistern or a drainage channel.

Vitruvius' descriptions are unambiguous, and clear enough for working models to be built. Some near contemporaries also refer to water power; their descriptions are far less clear, but they are good enough to convince scholars that there was something real behind their words. Lucretius (96–55BC) wrote in *De Rerum Naturae* of rivers turning wheels and buckets, possibly referring to the noria. Strabo's *Geography* of c.65BC makes reference to a watermill at the palace of Mithridates at Cabira in Pontus. The poet Antipater wrote an ode in which he urged the 'maids who toil at the mill' to cease their work, for Ceres had ordered the water nymphs to do the job of grinding for them. (There are several Antipaters to choose from, but most scholars take this to be Antipater of Thessalonica, c.85BC.)

Debate has centred on whether these writers were describing something new or whether they were making reference to an established technology that was sufficiently well known for assumptions to be made about the reader's understanding of it. In the absence of other unequivocal documents, caution has preferred the view that the waterwheel was new. There are earlier sources, in particular the work of Philo of Byzantium, whose writings date from about the 230s BC. In his *Pneumatics*, there is a description of an overshot waterwheel and the right-angled gearing that could drive a millstone. Philo also described a water-powered device, a whistling toy. There is another text by Apollonius of Perge, who, originally writing c.240BC, also referred to the 'newly invented vertical mill'. The problem with both the *Pneumatics* and Apollonius is that they survive only as later Arabic translations, and most scholars of the twentieth century believed that the texts were considerably amended, with additional descriptions of later devices. More recently, however, M. J. T. Lewis has come down in favour of the Arabs having made faithful translations, thus placing the origins of the watermill at least some two hundred years before Vitruvius.

The origins of water power may well remain 'enigmatic', as Orlan Wikander has described them, with debate continuing about detailed interpretations. Enough has emerged, however, to suggest that the waterwheel did come into use at some time during the two centuries or so before Christ.[9]

A further area of debate has centred on the type of waterwheel that was being described in the early texts. Vitruvius was clear: his was a vertical water-

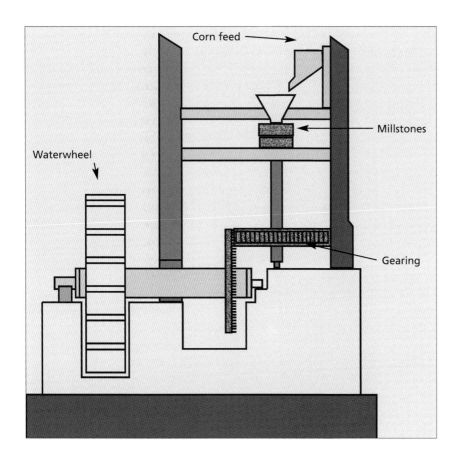

The basic layout of a watermill as described by Vitruvius. In all essentials it was the same as the medieval mill.

wheel, and he makes no mention of a horizontal type. Antipater's description, however, might be of a horizontal wheel. This was the interpretation preferred by Richard Bennett and John Elton in their multi-volume history of corn-milling, written at the end of the nineteenth century, and their lead has been followed by a number of more recent writers. This horizontal wheel, they believed, pre-dated the vertical by a considerable margin, the one leading to the other. Others, notably Terry Reynolds, are not convinced: the horizontal wheel of the Mediterranean region was probably devised more or less concurrently with, and independently of, the vertical. Convinced of its origins, Bennett and Elton referred to the horizontal wheel of southern Europe as the 'Greek mill', as distinct from the 'Norse mill' of northern and western Europe, and both names have stuck. Documentary sources for the Norse mill are lacking, but archaeological evidence is more abundant, revealing many watermills using the horizontal wheel in northern and western Europe, as far as Ireland, during the centuries up to the end of the Roman Empire. It does seem almost certain that the technology of the horizontal waterwheel was developed independently in northern Europe. How valid the distinction is between Norse and Greek mills remains open to debate.[10]

Vitruvius noted that the machines he was describing were ones that his readers might 'only infrequently have personal contact with', which suggests that the watermill was then uncommon, at least where his book would be read. That is consistent with a view that the technology of watermills, exploited first in Asia Minor, was only then beginning to be adopted in the centre of the Roman Empire. References to watermills are rare in Roman sources of the first two centuries AD, indicating that they were still uncommon. Various suggestions have been made to account for the slow up-take of water power. The supply of water is one possible factor.

Making good use of water power in such work as grinding grain requires a reasonable and steady supply of water. Throughout much of the Mediterranean world the rivers did not always flow with regularity. Pontus, where Strabo described one early mill, did have a secure catchment. However, this was not necessarily a long-term problem, for, once the principle of water power had been established, those who wanted to use it set about ensuring an adequate supply of water.[11]

Perhaps one of the more significant factors in the delay in any widespread adoption of water power was a plentiful supply of human labour, and indeed of animal power, in the early Roman Empire. Labour was provided either by slaves or was relatively cheap, and could be turned to other employment when not milling. Although there was unlikely to have been any incompatibility with employing the manual labour of slaves alongside a new mechanical device, landowners probably found the high capital cost of the watermill a serious consideration. It was especially risky at a time when the technology of waterwheel design and construction was still at an early stage, resulting in wheels that produced no more than 3–4hp and were not of the highest efficiency. The early users of waterwheels tended to be the larger estates, which worked at a greater scale and intensity.

It was not until the fourth century that watermills started to increase significantly in number. In that century Palladius, a writer on agriculture, advised landowners to install a waterwheel on their estate to save labour. What had been a surplus of workers was turning into a shortage in the later empire, and this probably lay behind the more rapid increase in the number of watermills towards the end of that period. By this time they could be found throughout the Roman territories. Archaeology has revealed a substantial number of sites, many in Britain; indeed, more sites have so far been identified in the British Isles than in any other country of northern Europe.[12]

EARLY WATERMILLS IN BRITAIN

The earliest date attributed to a British watermill is 150AD for a mill at Ickham, Kent, while two mills dating from the third century have been excavated at Chesters and at Haltwhistle Burn along Hadrian's Wall. They were among the first to be positively identified as watermills, and conforming to the Vitruvian type with vertical wheels. At Haltwhistle the wheel was probably about 10 feet in diameter. They are also the only mills of this antiquity so far discovered in northern England, whereas several Roman mills have been identified in southern and south-midland counties.[13]

Although a fair number of sites of watermills from the Roman period have been identified in Britain, the progress of the technology in the following centuries is not clear. It might in some places have survived the collapse of the empire. On the other hand it might have had to be reintroduced. The earliest post-Roman mill so far discovered is a tide mill at Nendrum in Northern Ireland, built in the early seventh century. Archaeological evidence for new watermills built in Saxon England before the seventh century is lacking. An early Saxon mill found at Old Windsor, Berkshire, dating from the late seventh century, had a vertical wheel fed by a channel from the Thames more than a mile long. At some time in the ninth or tenth centuries, this mill was replaced with another, using a horizontal wheel. This may be evidence that, at this time at least, the vertical wheel was not always regarded as representing progress. Horizontal wheels were probably more common in Saxon England than surviving archaeological evidence suggests. Among those that have been discovered was a Saxon watermill with horizontal wheel, revealed by excavations at the deserted medieval village of Wharram Percy in Yorkshire in the 1970s.[14]

There has been more plentiful archaeological evidence from Ireland, where nearly a hundred sites have been excavated revealing mills. Most were powered by horizontal wheels, but at least two examples dated to the eighth century show evidence of vertical wheels, demonstrating that the two technologies were contemporaneous in Ireland. The introduction of dendrochronology (tree-ring dating) in the 1970s enabled more accurate dates of construction to be ascribed to the mills, resulting in thirty-two identified as dating from before AD1000. This also identified a marked increase in the number of mills built during the eighth and ninth centuries, probably as a result of the increasing population in Ireland.[15]

One important find was a mill at Tamworth, Staffordshire, excavated in 1971 and dated to the middle of the ninth century. It was operated by a horizontal wheel fed by a leat cut from the River Anker and finds from the excavation have revealed some wood craftsmanship of high quality. By the ninth century, when this mill was built, documentary references to watermills were increasing. The first written evidence for a watermill in Britain appeared in a charter issued by King Ethelbert of Kent in 762. It recorded the exchange by the minster church of St Peter and St Paul, Canterbury, of a half share in the mill at Chart for some pasture rights in the Weald. During the following two centuries references to mills occur more often in charters, and in the early monastic chronicles. The various works associated with watermills, such as dams, weirs and watercourses, also receive mention, while place names indicative of mills also appear. By the tenth century it is clear that mills were not rarities. The construction of Abingdon Abbey mill is recorded at about 960, while elsewhere on the abbey's estates a mill at Tadmarton featured in a charter of 956. When the record of Domesday Book was compiled in 1086 the abbey had twenty-seven mills on demesne manors, and six more in the hands of tenants.[16]

3 The Medieval Mill

By the time of Domesday Book watermills were common enough to be regarded as a normal feature, a reflection of the extent to which water power had expanded during the late Saxon centuries. As one of the major studies of the survey remarked, 'A significant change had taken place in the material culture of England which removed the preparation of grain from the hearthside, altered the function of the housewife, introduced the miller and made the corn mill not only an important element in rural economy but a source of profit to those who erected them upon their holdings.'[17]

Domesday is one of the few sources where numbers of mills are quoted, and interpretation of the record is not easy. The surveyors were recording mills as the grinding machinery. Two mills in a manor did not always mean two separate buildings, although sometimes it did. The survey records assets of manors, and there were a number of different manorial rights applying to the mills. Some mills were shared between manors, accordingly being entered as a half mill under each manor. Professor H. C. Darby's best estimate, however, is that there were as many as 6,082 watermills in England at this time. His is the highest figure: there are others who would put the total at 500 or so fewer.[18] Many of these sites have been occupied almost continuously since that time, the mills being adapted and rebuilt over the centuries. In consequence, there is an almost complete lack of physical evidence of the mills in use at the Domesday period.

Watermills were certainly not in universal use, and their distribution recorded by Domesday Book was very uneven. There were only six in Cornwall, for example, very few in parts of Devonshire and Cheshire, and Staffordshire could muster no more than sixty-five. Lancashire apparently had none at all – whether that was so or whether there were some that went unrecorded is not known. In the north of England generally mills were more thinly distributed. Yorkshire, England's largest county, had only about a hundred. The limited extent of arable cultivation in the north, and the lower population, were two reasons. In addition, the region had only recently been the scene of William the Conqueror's campaign of 1069–70 to subject it to his rule. This had brought much destruction, which had not by any means been fully repaired by 1086. According to Domesday, a large part of the territory was 'waste', and that included the watermills. Recovery was certainly in progress and Domesday also records this fact, identifying locations where new villages were being built with a mill incorporated in their planning. Redevelopment was certainly starting in Sutton-under-Whitestoncliffe, a small settlement in north Yorkshire, as it was in the much larger township of Guisborough. Both had their own mills.[19]

At this period, mills were most common in the more settled districts of the Midlands, and eastern and southern England. Norfolk had the largest number recorded in Domesday Book, 538, while Kent, Hampshire, Wiltshire and Gloucestershire also had significant numbers. By relating the numbers of mills recorded to the numbers of tenants of the manors, an estimate of the distribution of mills to population can be made. Professor Darby calculated the average density for the whole of England to be twenty-five mills per 1,000 men. Those areas best served with mills appear to have been the southern counties, from Kent to Dorset, Wiltshire and Gloucestershire. With one mill to every twenty-five to forty households, these counties were more

The Medieval Mill

The mill that stands at Houghton in Huntingdonshire dates from the 1660s, but it is in a place recorded in Domesday Book. (Museum of English Rural Life)

densely provided for than Norfolk, which, although it had the largest number of mills in the country, actually had one mill to every fifty households.[20]

Domesday Book placed a value on the mills. Those with the highest value were clustered around main centres of population and located on the bigger rivers, which had an assured flow of water. At the other extreme were the small mills, given a low valuation, on the smaller streams. The low population and limited agriculture of the Tame valley in Staffordshire could support only five small mills. The values recorded for them were no greater than 2 shillings, a fifth of the amount of the mill at Brailes in neighbouring Warwickshire. Many of these low-value mills worked only part of the time. There are many references to 'winter mills', used only seasonally when there was sufficient water. Not all the mills recorded were in use at the time of the survey, and such mills were entered as being either disused or decayed.[21]

In the southern counties there were already some local areas that had quite a dense distribution of mills. Some rivers, such as the Wylie in Wiltshire and the Loddon in Berkshire, supported numbers of mills not far short of those of the eighteenth or nineteenth centuries. There were also a few distinct clusters of mills: nine were recorded at Old Alresford, Hampshire, and seven each at Downton, Pewsey and Warminster in Wiltshire.

Few, if any, of the mills identified in Domesday Book were employed for purposes other than grinding grain. The record of two mills making rent payment in iron has prompted the thought that they might have been ironworks using water power. If this is indeed what they were – and it is not certain – they were rare exceptions for eleventh-century England. Corn-milling predominated at this time and, even after the adoption of water power for other industries, continued to do so throughout the medieval period.

A growing population increased the demand for flour, and with it the need for powered milling during the two centuries following the Domesday survey, although grinding by hand quern remained common. By 1300 the population of England had more than doubled, from about 2–2.5 million to 5–6 million. The number of mills had grown also: there were probably more than 10,000 in England, and perhaps as many as 12,000. Many of the new ones were windmills, exploiting a technology introduced in the late twelfth century. Windmills were cheaper to construct, and could be built in places where the water supply was limited, such as the Fens of eastern England. Nevertheless, the number of watermills had risen to 8,000–9,000 by 1300. Mills had grown bigger as well, with rationalization taking place in some manors, to concentrate on a few large mills on the major streams with guaranteed water. Some smaller mills were closed, resulting in a reduction in the number of mills in some manors, as at Blockley in Gloucestershire, where twelve mills became three. In the county of Huntingdonshire there was a decline in the total number of mills between Domesday and the survey for the Hundred Rolls, in 1279, but on the major rivers of the Nene and Great Ouse numbers had increased. All of this was happening at a time when there was an increasing demand for water power, as it was gradually being introduced to other contexts, such as the textile, tanning and metal industries.

Growth in the number of mills was sent into sharp reverse following the famines and epidemics of the Black Death in the fourteenth century. The population of England was halved in a few years after 1350, returning to the 2.5 million or so of the late eleventh century. With a smaller population to feed, land was turned from arable to pasture, fewer cereals were grown, and grain-milling declined. The manorial revenues from mills fell; mills were left idle for long periods; some were converted to other uses, perhaps less profitable than corn-milling; eventually, some were closed. The evidence is everywhere to be found. In the 1320s the estates of the bishopric of Norwich reported that they had fifteen watermills and twelve windmills. In 1369 the totals had fallen to ten water and six wind. The mill at Cookham in Berkshire had been valued at £4 in the 1320s. In 1370 its value had fallen to £2 6s 8d, and by 1381 it was disused and in disrepair. At Newbury the tanning mill was described as having no value after the Black Death. Domesday Book had recorded five mills at Standon in Hertfordshire, where the river provided a good source of power. All were corn mills, together worth £2 5s. By the 1330s, after the famines, one of them had been turned into a fulling mill, but was yielding little for the manor; the other four, still corn mills, together were valued at £1 6s 8d. After the Black Death matters deteriorated further, and another of the mills was converted to fulling at a much lower rent. There was a short-lived recovery in fortunes in the late fourteenth century, followed by further decline in the first half of the fifteenth, during which time one of the mills ceased work, leaving two corn mills and two fulling mills. The rent of Great Shelford mill, Cambridgeshire, which had been £10 in 1322, was £6 6s 8d in the 1350s. This site did recover slowly thereafter, until by 1469 its rent had risen to £11.[22]

The medieval mill was a manorial right, the 'milling soke', and from this, the manorial mill became known as the 'soke mill'. The lord of the manor had sole claim to build a mill. Effectively there was one per manor: Wharram Percy was one place where there were two manors in medieval times, and there were two mills in place of the single Saxon one. Amalgamation of the manors in 1254 led to one being taken out of use, and to its millpond becoming a fishpond. Having the only mill, the lord could then compel the tenants who held land under feudal (or customary) tenure to use it, a practice known as 'suit of the mill'. All corn grown and consumed in the manor had to be ground at the lord's mill. Even grinding a little grain at home with a hand mill could be an offence. There was a similar practice in Scotland, which was known as 'thirlage'. Free tenants were not under the same obligations, but of necessity used the manorial mill as the only one available.

Suit of the mill was a valuable right, which most manorial lords were keen to uphold. It arose as a result of those aspects of water power that had been constant throughout its history: the high capital cost and relatively low running costs. Having made a substantial capital investment in his mill, the manorial lord was usually keen to protect it and to recoup the money. Giving judgment in a Scottish thirlage case in the eighteenth century, Lord Kames held the view that 'it was thought to require the privilege of a monopoly to encourage men to lay out their money upon works so generally useful'. The claim that the lord was providing a valuable service for his tenantry was one sometimes put forward, but the tenants were not always grateful, and many preferred the opportunity to go elsewhere.[23]

The medieval miller did not buy grain from the farmer, grind it and sell the flour. Indeed, he was forbidden to deal in corn and meal. Instead, he milled the grain brought to him by the manor's tenants; what they brought to the mill they took away as flour, less the toll. The miller was, therefore, working for the most part on the small scale, a sack or two at a time, and having to make sure that each grower's grain was kept separate. Operating costs were kept low by this arrangement, for he did not need many staff, as the farmer could be responsible for supervising the milling of his grain. No more than two men were needed at many mills until well after the medieval period. Nor did the miller need large amounts of storage space for grain and flour, and in consequence most mills were small buildings.

Payment for using the manorial mill was principally by way of a toll known as 'multure', whereby the miller took a proportion of the grain. This deduction was made by weight or by measure, with custom varying locally. There was also great variance in the proportion demanded. An English statute of 1270 said that the fee should be not more than one-twentieth or one-twenty-fourth, depending on local custom and the strength of the watercourse. In practice, multure ranged from one-thirteenth, which was usual in Durham and Northumberland, to one-thirty-second. These were the rates paid by customary tenants; free tenants usually paid less.

Additional obligations that could be imposed on tenants included duties to be performed towards the maintenance of the mill, in particular keeping the dams, ponds and watercourses clean and in good repair. In many places these rights were still being upheld late in the Middle Ages, as an enquiry into the manor at Norwell, Nottinghamshire, in 1406 demonstrates: 'Each tenant holding one bovate or a messuage in lieu of a bovate shall help cleanse the Lord's millpond for a day with one man, taking nothing for the work.'[24]

Disputes and prosecutions relating to manorial rights were many: tenants were brought before manorial courts for failure to keep suit of mill, whether for using a hand mill or for going to a mill outside the manor. One of the more celebrated examples was the long-running dispute between the abbey of St Albans and its tenants, which lasted for the best part of a century. The monopoly of the mill became part of a wide-ranging struggle over manorial rights, included as one of the local grievances during the Peasants' Revolt of 1381. One of the abbots paved a courtyard with querns confiscated from his tenants as he tried to cajole them into using the manorial mill.[25] As well as demonstrating the desire of lords, lay and ecclesiastical, to enforce their manorial rights, this story shows equally a spirit of independence on the part of the tenants. It also indicates that many of them did not need the services of a mill. The amount of grinding work they had was small enough to be managed with a hand quern.

Medieval mills might be directly managed as part of the manorial estate or let to an independent miller. From the miller's point of view, a tenancy was almost certainly preferable. The miller at the time of Domesday Book had servile status: he was on a par with the feudal peasant. Whether formally stated or not, this status continued for the next three centuries or so. As a consequence, wages paid to the miller employed by the estate were very low, with no allowance being made for his skills. In monetary terms the wage was often less than 6s a year in the thirteenth century, although he might get some small share of the tolls paid. Out of this he usually had to pay the boy who was his assistant.

The tenant of a mill was the one who had some prospect of earning a decent living. Indeed, given the right situation and trade, he could become reasonably well off. Rents paid for mills depended on their trade, and ranged widely. In Domesday Book some mills were valued as low as 3d or 4d; higher up the scale, there were three mills in Cambridgeshire with values of £3. There were many non-monetary rents as well, most of them paid in wheat, although flour, malt, eels (fish), salt and iron are also mentioned. This type of rent was still common in the thirteenth century: '16s and 20 sticks of eels' was the rent for Sturminster Newton mill, Dorset. Later in the Middle Ages, rents settled down to be mainly quoted in cash terms. By the fourteenth century these ranged from £2 to £5 a year. The range of values continued beyond the medieval period and in the seventeenth century rents were generally between £5 and £15 a year.[26]

The letting of mills ('farming' was the common medieval term) was more regularly practised by the thirteenth century. Large estates, especially monastic ones, favoured this management for outlying mills. After the Black Death leases became more common and it was not always the working miller who took the lease from the manorial estate. Other businessmen leased mills as investments, and then either sub-let to a working miller or directly employed him. With this, 'miller' became a less specific title, and this remained more or less the case; a 'miller' could now be either the man who leased the mill or the one who worked it.[27]

The monasteries were among the major medieval landowners and users of water power. They varied in the way they managed their mills – many monastic estates were extensive, and these often acted like manorial lords, letting out at least some of their mills. St Albans Abbey was one example of such a system. The needs of the abbeys themselves for flour, meal and ground malt were considerable, and to meet these they operated as self-sufficient houses.

The Medieval Mill

When Frank Gregory saw the medieval mill building at Fountains Abbey in 1963 it was still derelict awaiting restoration. (The Mills Archive Trust)

Monasteries sought sites that could accommodate mills built within their precincts, and these mills were directly managed by a paid servant or a lay brother of the house. A large monastic house was often prepared to invest in its mills. Buildings were substantial, often of stone (others were timber-framed), and with more storage capacity than the manorial soke mill. Watercourses were designed both to feed the corn mill and to contribute to the domestic water supply and drainage of the abbey. Originating in the twelfth century, the building that survives at Fountains Abbey housed two mills fed by a millstream from the River Skell. As well as the corn-milling, monastic estates were often at the forefront in introducing water power for other industrial applications. Fulling mills in particular were often built by the abbeys, which had extensive sheep and wool enterprises.

The decline of manorial rights during the late medieval period and on through the sixteenth and seventeenth centuries by and large freed people from their manorial obligations, at least in southern England. Independent mills grew in number, and the selling of grain off the farm to millers or dealers also increased. Some manorial rights did survive in the soke mills until the late eighteenth century in parts of Wales, the west country and northern England. The last one in Derbyshire, reported John Farey in 1806, was at Castleton. In Ireland, also, it was common at the end of the eighteenth century to see a clause in a lease stipulating that the tenant must grind his grain in the lord's mill. By then, however, it was almost impossible to enforce. In Scotland an eighteenth-century campaign culminated in the Thirlage Act, passed in 1799, abolishing this remnant of the feudal system. In some towns of northern England, such as Leeds, Bradford and Malton, the manorial mills were kept up until the mid-nineteenth century. The most notable were the soke mills of Wakefield, a town with one of the largest grain markets in England. It was not until 1853 that the inhabitants of the town were able to redeem the manorial rights, paying £18,000 for them. The soke rights at Leeds had been similarly bought out in 1839.[28]

4 The Technology of Water Power

The watermill is a complex mechanism. The width and diameter of the wheel, the gearing and bearings, the speed and flow of the stream and the fall that can be achieved are some of the engineering and hydraulic issues that have to be faced by its builder.

Since these engineering conundrums must be tackled, water power, although a readily available, 'green' resource, has never been free. The costs of the dams, watercourses, wheels, gearing and buildings have always been significant. However, the value of the power was such that builders of mills were often prepared to go to great lengths to harness it. There have always been alternatives to water power, and it was because windmills were often cheaper that, almost from their introduction, they provided competition that was sufficiently effective to replace some watermills.[29] There are limitations on the amount of power that can be obtained from a mill, and much of the technological development over the centuries has been aimed at increasing efficiency and productivity.

WATERWHEELS

The waterwheel is at the core of water power, the prime driver and source of motive power. As demand for power increased, it is not surprising, therefore, that much attention was focused on the efficiency of the wheel.

Waterwheels can operate in either the horizontal or the vertical plane. Although it is usual now to think of a mill wheel as vertical, the horizontal wheel has at times been at least as important, and can still be found. Probably because it is a relatively simple mechanism and the vertical wheel became the more common, there has been a tendency to assume that the horizontal wheel was introduced some time before the vertical wheel. The simple does not necessarily always come first, however, and there are those who would agree with Dr Lewis that the two types of wheel were probably developed concurrently, in the Hellenistic world of Byzantium and Alexandria, and that they were taken into use early on mainly in Asia Minor.[30]

Horizontal wheels work by the motion of water against vanes or blades pushing them round, and are thus known as 'impulse' wheels. The blades vary in shape: they might be flat, and set straight to the axle or at an angle, or they might be curved. They usually rotate at quite high speed, and need a fall of about 10 to 12 feet. A horizontal wheel can be quite small and in its simplest version may be planted straight in the flow of a stream, without the need for millstreams, ponds, sluices and so on. A description of such a mill appears in a seventeenth-century document from County Down, referring to its driving small quern-sized stones.[31] However, it was usually found best to construct some diversion of water, with a pond to store the water, and a wooden trough to direct the water to the wheel at a suitable angle and with sufficient volume and velocity.

The mechanism from a horizontal wheel is also simple: a vertical shaft from the wheel gives direct drive to a pair of millstones without the need for arrangements of gearing. In the most common arrangement for this type of mill, the water is directed down the shoot (penstock) so that it hits the wheel with sufficient force to turn it (*see* overleaf). The building is of two storeys, the lower one the wheel-house and the upper the stone floor. At the top

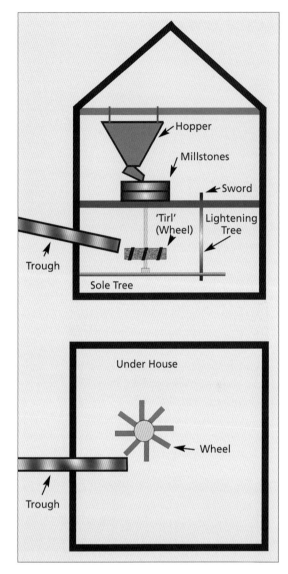

The form of operation of a horizontal mill wheel.

of the vertical shaft from the waterwheel an iron spindle, the 'rynd', was inserted. This had a 'T' section and was fitted into the upper millstone, enabling the power from the wheel to turn the stone. Another shaft, called the 'lightening tree', passing from wheel-house to stone floor, allowed the gap between the stones to be adjusted. The miller could work a lever at the top to raise or lower the lightening tree, which in turn altered the height of the sole tree, the platform for the waterwheel. That had the effect of moving the waterwheel up or down, and at the end of this cycle the upper millstone was lifted or lowered.

With its direct drive from wheel to millstone, the mill with a horizontal wheel was generally simple and cheap to construct and maintain. It was work that could be undertaken by the small-scale and peasant operator. A small mill of this type was to some extent portable, and could be removed to a better stream if need be. With its smaller wheels and their speed of rotation, the horizontal type was suited to small, fast-flowing highland streams; preserved examples can still be found in the Shetlands.

The horizontal wheel of medieval times was far less efficient than a vertical wheel, using between 5 and 15 per cent of the available energy compared with the 30 per cent of a vertical undershot wheel. Without the intermediate gearing it was less able to maximize output; through its gearing, the vertical wheel offered greater control over the speed at which the millstones operated. These factors probably explain why by the eleventh century the larger manorial mills had already opted for the vertical wheel.

It is tempting to think of the horizontal wheel as primitive, a type superseded by the vertical wheel, and this is certainly how British engineers and a good many historians have tended to view it. They have been encouraged in such a view by the disappearance of the horizontal wheel from England quite early in the Middle Ages. Surviving examples were confined to the peripheral parts of the British Isles, including some areas of Ireland, and Shetland, where they continued in use well into the nineteenth century. The first edition of the Ordnance Survey six-inch map in 1878 recorded as many as 500 mills with horizontal wheels. They were small and seasonal, for ponds and other means of controlling the supply of water were rarely added to the site. Archaeological evidence for horizontal wheels was also more abundant in these regions.

In England, although certainly there were some mills with horizontal wheels in Saxon times, there is little to show of any of them. It has been suggested that some of the small mills recorded in Domesday Book were powered by a horizontal wheel. Neither

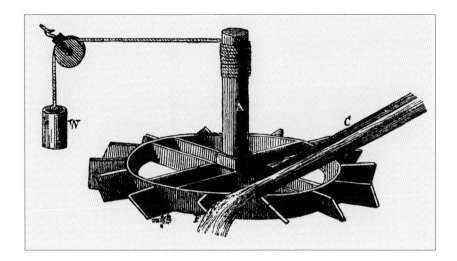

A horizontal wheel illustrated in Tomlinson's Cyclopaedia of Useful Arts *(1851). (Museum of English Rural Life)*

the Domesday record nor other evidence confirms that, but there may have been some at work then. The vertical wheel was dominant, and Professor Holt attributes this to the fact that most mills were part of manorial or monastic estates. The lords were able to invest in bigger, more efficient mills, for which the vertical wheel was then the best option. The cheaper horizontal wheel was for the peasant, and this became an assumption regardless of any other merit the horizontal wheel might have had. The vertical wheel did, indeed, supersede the horizontal one as manorial and commercial mills achieved dominance.

The process is better documented in Scotland, where the change took longer. It is clear that peasants, independent farmers and crofters were the ones who used horizontal wheels, while the lord built the bigger, vertical wheels. Progressive opinion in Scotland was always contemptuous of the little horizontal mills, which were described in terms such as 'wretched', 'miserable' and 'trifling'.[32] The uses to which the horizontal wheel was put largely bore out that opinion. It was a small, simple device, often roughly made from materials available locally, driving small stones that were also hewn from local supplies. The mill building had to straddle the stream so that the direct drive shaft from the wheel could come up through the floor, and this made it vulnerable in times of spate. The crofter, however, was happy with this type of mill; the wheel might have been inefficient, but it gave him the power he needed at minimal cost.

Because of its lowly associations, the horizontal wheel remained largely undeveloped in the British Isles, while in small-scale Scottish grain-milling it continued in limited use for a long time. The last such wheel in mainland Scotland worked at Kinlochbervie until 1864. Otherwise, most were in the Western Isles, Shetland and Orkney, where some worked into the twentieth century. Attempts to do more with the horizontal wheel were rare. There was said to have been a horizontal wheel at a seventeenth-century paper mill at Restalcraig on the Tumble River, although the evidence is insubstantial, based mainly on a street name surviving to the present day.[33] There is firmer evidence that John Smeaton did install one in a corn mill at Scremerston, Northumberland, in 1776. Its mode of working was similar to that of an axial flow impulse turbine and at the time it was unique in England. In continental Europe the vertical wheel had achieved similar dominance during the Middle Ages, but in the sixteenth century interest in the horizontal wheel revived. Work was done to improve its efficiency, and its use began to increase, not only for the small-scale peasant operation. Technical development continued from the sixteenth century onwards, establishing many of the principles that were followed by later water turbine technology.

Whatever the differences in sophistication

The Technology of Water Power

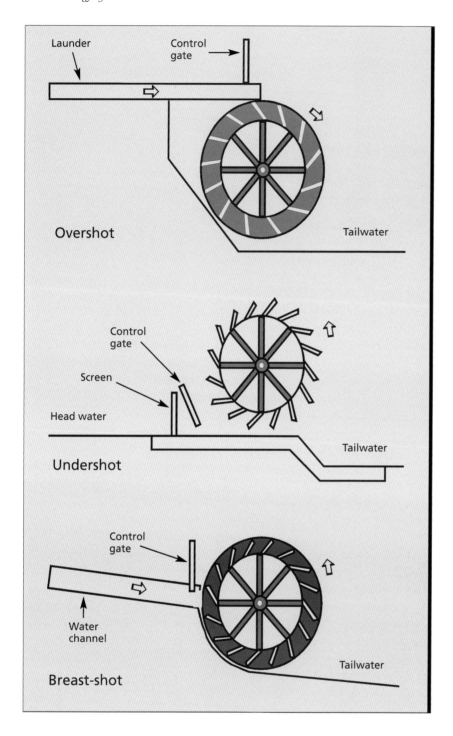

The different types of vertical waterwheel.

between horizontal and vertical wheels, there is no doubt that the vertical wheel had become the dominant type throughout most of Britain by the end of the Middle Ages. The vertical waterwheel drives a horizontal shaft. There are three different forms, dependent on the position at which the water strikes the wheel: undershot, overshot and breast-shot, the naming derived from the fact that water is directed or 'shot' on to the wheel (*see* above).

The undershot wheel dips into the stream to be

struck at the lowest point of the revolution – at 6 o'clock. It is relatively simple to construct, with the flow of the water against paddles ('float boards') pushing the wheel round in an anti-clockwise direction. The overshot wheel is turned by the water being dropped on to the upper part of the wheel, usually at about the 1 o'clock point. Whereas the undershot wheel is driven by the force of water on its blades, the overshot wheel is a gravity machine. The water falls into compartments ('buckets') on the wheel, and the weight of the water then forces the wheel to turn in a clockwise direction, opposite to the travel of the undershot wheel. To achieve that fall of water usually requires the construction of a channel leading above the wheel, and a wheel that is strong enough to take the weight of the water. It has on the whole, therefore, been more expensive to construct an overshot wheel and the associated water channels. The wheel was usually built into a pit from which the millstream flowed out after turning the wheel. The engineering did have considerable benefits – having an artificial mill leat bringing the water allowed for greater control over the flow to the wheel. The overshot is the more efficient type, a

An undershot waterwheel at Mapledurham Mill on the River Thames.

Dunster Mill, Somerset, had twin overshot wheels. Looking the worse for wear in this photograph from the early 1950s, they have been restored to operational use. (Museum of English Rural Life)

The Technology of Water Power

matter conclusively demonstrated in the eighteenth century by John Smeaton. It was clearly appreciated in the medieval world, and was employed wherever possible, in the iron industry, for example, where the extra power was very useful.

The third type of vertical wheel is the breast-shot wheel, in which the water is dropped on to the floats half-way up the near side of wheel, the force turning the water in an anti-clockwise direction. It is so called because of the masonry 'breast' built up to surround the descent of the wheel. Essentially, this was the wheel to have where conditions were not suitable for an overshot wheel. Depending on the point at which the water hit the wheel, it was referred to as a low breast-shot, breast-shot (hitting the mid-way, or 9 o' clock point) or high breast-shot.

The undershot wheel was the earliest type of vertical wheel, the one described by Vitruvius in 25 BC. It was the most common type, certainly until the eighteenth century. The overshot wheel seems to have been known from late Roman times, but its use was probably limited. Illustrations of waterwheels in medieval documents do not show overshot wheels before the fourteenth century. The Luttrell Psalter of this time indicates its presence in medieval Britain. This would suggest that the overshot wheel was increasing in use, but the extent of its use is impossible to determine. It has been suggested that in medieval times the overshot wheel was used where water supply was difficult, the efficiency of the wheel squeezing out the most from the available water. Some of the medieval overshot wheels seem to have been constructed with floats rather than buckets, making them more impulse engines than gravity.

On wheels with buckets, most were of simple design, effectively floats curved upwards or set at an angle to catch the water, and enclosed within the rim of the wheel. Spillage as the bucket descended, with water impeding the wheel and consequent loss of power, was a serious problem with these buckets. A bucket constructed as a bracket from the hub, known as the elbow bucket, held the water better, giving improved power and efficiency. Elbow buckets were introduced by the early fifteenth century, but were not in common use before the eighteenth.

The breast-shot wheel was to become important

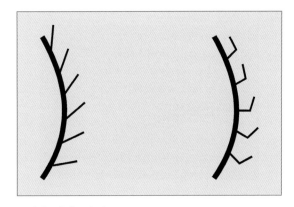

Angled and elbow buckets.

in later centuries, but in the Middle Ages there is no evidence of it. The first description of such a wheel seems to be by Leonardo da Vinci in the late fifteenth century. John Fitzherbert's *Boke of Surveying* indicated that he, too, was aware of it in the sixteenth century.[34]

Wheels were constructed of wood and built up in two ways. One resembled a wagon wheel, with spokes mortised into the hub and the rim in sections (or 'felloes'). This type is often referred to as a 'compass-arm wheel'. The second was the 'clasp-arm wheel', on which the hub was squared off and gripped by pairs of parallel spokes, one of the pair on either side, attaching it to the rim. This had the advantage that there were no mortises on the hub into which water might seep. Its weakness was at the fixing of the spokes to the rim. To judge from the small number of contemporary illustrations showing them, clasp-arm wheels were less common in Britain than they were in mainland Europe, from where there are many depictions. However, this judgement is probably misleading, and clasp-arm construction is likely to have been more common in Britain than the number of illustrations would imply. John K. Harrison has found evidence of some clasp-arm waterwheels in north-east Yorkshire, a few of which survive. One at Coulton Mill has an unusual arrangement of crossing arms of two-thirds length, which forms a Star of David pattern round the hub.[35] These examples are from small rural mills and it is clear that the Industrial Revolution edged out clasp-arm construction for waterwheels. Even then, some pitwheels and spur wheels continued to be made in this way.

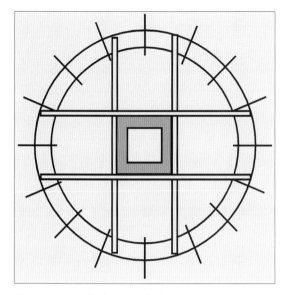

The construction of a clasp-arm wheel.

The waterwheel illustrated in the Luttrell Psalter is shown with five spokes. The evidence of archaeological excavation also suggests that wheels of the medieval and early modern period were often of no more than six or eight spokes, a relatively light construction.

The size of wheel was determined by a number of factors, including the use to which the wheel was to be put, the size of stream and the head of water that could be obtained. Local terrain therefore played a part. Wheels in low-lying territory were, on the whole, likely to be smaller than those located in places where a good fall could easily be achieved. The River Tame in Staffordshire descended by only 300 feet in 20 miles, at an average gradient of 1 in 475. Obtaining much of a fall to the wheel was not easy, and the largest-diameter wheel on this stretch was 17 feet. Likewise in Suffolk, when the Society for the Protection of Ancient Buildings conducted a survey after the Second World War, the largest-diameter wheel found was 20 feet 8 inches and the smallest 8 feet. Some users of water required a smaller wheel, some needed a bigger one. On the whole, a fulling mill needed a smaller-diameter wheel than a corn mill. Measurement of a number of wheel-pits from seventeenth-century mills in Ulster showed the average size for corn mills to be 14 feet 6 inches,

Two large-diameter wheels in an overgrown state at Gweek, near Helston, Cornwall. (Museum of English Rural Life)

compared with 12 feet for fulling mills. At the other end of the scale, mining was the principal user of large wheels.[36]

Information about the size of waterwheels before the Industrial Revolution is not plentiful. Wheels in the Roman world were probably of about 6–12 feet in diameter.[37] In the early modern period they were perhaps 10–15 feet. It is not until the eighteenth century that writers start to mention the dimensions of wheels with a regularity sufficient to gain an impression of how big the wooden waterwheels might be. Engineers such as John Rennie and John Desaguliers record a number of wheels in Britain with diameters ranging from 12 to 18 feet. Most were undershot wheels employed in fulling mills, paper mills and in the iron industry. Corn mills are rarely mentioned, but they might possibly have brought the average size down. There were some wheels of larger diameter, the best known of which was located at the London Bridge waterworks. It was 20 feet in diameter. There were some wheels of over 30 feet in the Cornish tin mines, and John Smeaton reported some of up to 48 feet in the 1750s. Those, however, were exceptions, for, as Smeaton noted, wooden wheels of more than 40 feet became 'heavy, unwieldy and expensive'.[38]

Where there is any information, indications are that most of the waterwheels before the Industrial Revolution were fairly narrow, between 1 and 5 feet. Again, there were exceptions: the London Bridge wheels, four of them in all, were each 14 feet wide.

Estimating the power of watermills before 1750 is no easier. The many variables for which figures are not available include the diameter and width of the wheel, its speed in revolutions per minute, the head and fall of the water flowing into it, and the width of the channel of water, both flowing in and the tailrace. Those who have tried to put figures to the power output by waterwheels have made widely differing estimates. Some have suggested very low power outputs from medieval and early modern wheels: Paul Wilson thought many produced less than 1hp, while Carlo M. Cipolla put the figure at between 1 and 3.5hp. Other estimates have been in the range from 2 to 7.5hp, but Leslie Aitchison thought that wheels of the fourteenth century could generate up to 60hp. Terry Reynolds, analysing the figures recorded for forty wheels across the whole of Europe in the eighteenth century, has concluded that the average power output was 6.5–7.5hp.[39]

A corn mill requires a modest amount of power. Boulton & Watt at the end of the eighteenth century reckoned that a pair of millstones could grind eight bushels of wheat an hour, and this would require 8hp. They later revised their estimates down to six bushels per hour. They were supported in their view by other engineers, such as Rennie and Sutcliffe. From insurance assessments of the period 1816–20, the average power of water corn mills in England and Wales seems to have been 14.5hp. Over 70 per cent of the mills, those with no more than two or three pairs of stones, were generating less than 12hp. By that date the average power was increased by the inclusion of some of the big wheels being built to meet demand from an industrialized society needing more power.[40]

DAMS, WEIRS AND WATERCOURSES

A watermill cannot, of course, work without water. Water supply and constancy of flow have, therefore, been a constant concern for millwrights, engineers and mill owners. Drought has always been a serious problem, reducing output or stopping it altogether, sometimes for months on end. Conversely, flood could be just as serious, even dangerous, and likely to damage the weir, wheel and structure of the mill. Less seriously, perhaps, winter frost and ice could also cause problems and stop the wheels turning.

Finding a good site was, therefore, obviously a first step towards a successful mill. With rivers still relatively empty, the medieval lord would often have his mill built at the downstream boundary of the manor so that it would gain from the greatest extent of water flow within his estate. There are even some examples of parish boundaries being moved to take in the mill. The mill was often sited just upstream of a ford, the nearby river crossing aiding access for the mill's users. The first site to be developed was not always the best. There are many examples of medieval mill sites that were discovered to have

The Technology of Water Power

Eversley New Mill is built alongside a ford.

insufficient water supply, and which were abandoned early on. One of these was at Wickwar, Gloucestershire, which had insufficient water for summer working and had ceased to be used by the mid-sixteenth century.[41]

The simplest way of taking up power was to place the wheel in the stream. This was a not uncommon arrangement with horizontal or undershot wheels. A high proportion of medieval mills were in the mainstream, and not a few in modern times. The mill was sited on the riverbank, sometimes straddling the stream. Where possible, a mill might take advantage of a waterfall acting as a natural dam and building up a head of water. There are, however, limits to what can be achieved with a wheel in the mainstream without any engineering intervention because the miller would have very limited control over the flow of water to the wheel. Building up the embankments of the stream could help to channel the full flow of water towards the mill, preventing it spreading out across the fields at times of flood, but this, too, was of limited effect.

Therefore, even a small horizontal wheel on a mountain stream or a mill under a waterfall would benefit from some further engineering to control the flow of water and create a greater head. This could be achieved by building a weir across the stream. It embanked the water above the mill, which was still positioned in the mainstream. A bypass channel would then be needed to take overflow water around the mill.

Mills in the mainstream could not usually be built on the larger rivers, where the flow might be too sluggish to provide power. One way to resolve that problem was to have a floating mill, sited on a boat in the river. The hull of the boat acted as a channel, funnelling the stream to the wheel. Floating mills are known to have been used when Rome was under siege in AD 536–7; later, many medieval European cities employed them, and they continued to be used

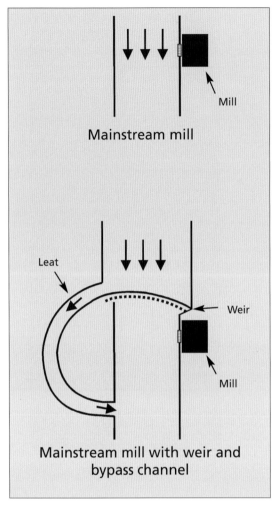

Some ways in which mills could be built in the mainstream of the river.

Taking the mill away from the mainstream became the most common way to improve the supply of water and overcome deficiencies in the natural flow on the river. Artificial cuts were then built, diverting water to serve the mill. Millponds and reservoirs could also be constructed to maintain the supply, and weirs and dams would divert water from the main channel. The introduction of weirs and their associated water channels was of major importance, Terry Reynolds has argued, for it had the effect of freeing water-powered industries from the constraints of siting. Streams were dammed for purposes other than watermills. Causeways were built to create foot-crossings; some ancient causeways were later adapted as mill weirs. There were many medieval fish-weirs and dams to aid navigation. What is regarded as the first true dam in England was built at Alresford in 1189 to improve the navigation of the River Itchen.[43] Until the late eighteenth century, however, the most common reason for building a weir or reservoir was for water power. Thereafter, canals and urban water supplies gave added impetus to the construction of other reservoirs.

A weir is a low dam (an overflow dam), which allows excess water to escape by flowing straight over the top, giving the impression of an artificial waterfall. Weirs for mills have often been referred to as hydro-power dams, although the term has also been used for the much larger dams built to serve twentieth-century hydro-electric power stations. There were three main reasons for damming the stream. One was to divert water from the stream to an artificial side channel, the millrace, leading to the mill. The second was to create a store of water, either by backing the water up in the stream behind the weir or by creating a separate reservoir, the millpond (or pound). The third reason was to build up the head of water, thus increasing the fall on the wheel. Quite often, one weir served more than one purpose. In all of these ways, damming the stream enabled mills to be built at more convenient sites, and with a more efficient supply of power. The result of centuries of activity by mill owners meant that most rivers came to be punctuated by weirs and networks of channels along their length. Many of the weirs can still be

in some parts of eastern Europe until quite recently.

Another way of using the flow of a large river was to site a mill under or against a bridge, the abutments and arches acting as channels, directing the flow to the wheel. Paris and Corbeil in France and a number of other European cities had bridge mills. Floating mills and bridge mills were rarely used in Britain, although there were some floating mills on the Thames in London for a few years in the sixteenth century, and again at the end of the eighteenth. The successful use of waterwheels on the Thames through London was for another purpose: four large wheels were built under London Bridge to pump water for public supply.[42]

The Technology of Water Power

A weir at New Mills, Cheshire. It has a pronounced top lip, and a gentle curve against the current.

The weir across the River Derwent at Chatsworth. Water was diverted above it through an underground channel to a corn mill about a hundred yards downstream.

The Technology of Water Power

The weir facing downstream at Masson Mills, Derbyshire.

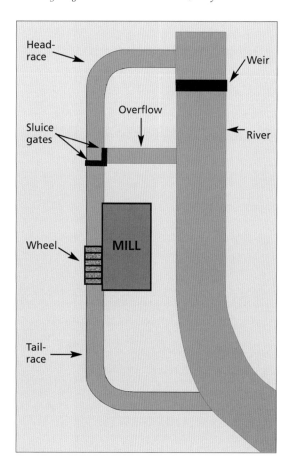

seen, and the channels traced, although the mills they served might have gone.

Weirs and millpond dams were of a simple construction; known as 'gravity' dams, they used the weight of the structure to hold back the pressure of the water. They had the disadvantage of being liable to weaken at any point, and indeed dams did give way on many an occasion. To overcome this, in the absence of the detailed knowledge of stresses that engineers later acquired, the tendency was to over-engineer. Earth, timber, rubble and stone were the main construction materials; where it was available in quantity, some were built entirely of stone.[44]

Weirs varied in their design and siting. Some went straight across the stream. More often, the weir was set at an angle to the flow of the river. In one common type, at least in later centuries, the weir had the form of a convex curve facing upstream; the weir on the Derwent in Derbyshire that served the Masson Mills was a rarity with its curve facing downstream.

(Left:) *A basic arrangement of water engineering for a mill on a leat off the river.*

The Technology of Water Power

This weir at New Mills diverted water through the sluice gate along a short channel to the wheel in the tunnel at the right of the photograph.

In its simplest version, the weir might divert the water a few yards along a channel to feed the mill just off the mainstream. Frequently, however, the weir had to divert water into a longer millrace of more substantial construction.

The need for greater control over the supply and flow of water, the desire to gain extra power, and the effects of building more mills along the same stream, all contributed to the need for artificial watercourses to feed the mill and take away the spent water. The millrace ran alongside the mainstream, often at a gentler gradient. This enabled the head to be built up above the wheel. The section of the millrace above the wheel was known as the 'leat' (or 'lead'), while the outflow channel below the wheel was the tailrace.

The channel to carry the water above overshot, pitch-back and many breast-shot wheels was known as a 'launder' or 'pentrough'. It was usually constructed as a wooden trough, but in later years iron was often used. Most were only a few feet in length, but sometimes they were much longer, and might carry the water over dips in the ground, roads and tracks. At one early seventeenth-century blast furnace in the Forest of Dean a channel 75 yards in length built from tree trunks brought the water to the wheel.[45]

Some mills drew their water straight from the leat. Others had a millpond to act as a storage reservoir and to build up the head of water. In upland areas there were mills with ponds fed from nearby springs, and others for which a pond was dug out to gather groundwater from the neighbouring hills. There were some medieval examples in north Yorkshire. On the whole, these ponds were able to gather sufficient water only for the small and part-time mills. Other mills needed a more reliable system that would give the miller greater control. In this situation, a pond would be built that would be fed from the stream, or on or adjacent to the millrace. Some large reservoirs were being built in medieval times, including those that served Byland Abbey and

The Technology of Water Power

A launder carrying water to an overshot wheel at Synod Mill, Cardiganshire. (Museum of English Rural Life)

Newland Priory in Yorkshire. This involved the diversion of streams a considerable distance to feed a series of ponds.[46] Millponds were often held back by a dam proper, one for which the overflow is usually an escape channel for surplus water, not over the top of the wall. They were built as close to the mill as possible, preferably immediately above it. Circumstances did not always allow that, and quite long channels might be necessary to lead the water from millpond to wheel.

The waterwheel was set turning by the release from the leat of a sluice gate, known as a 'penstock'. Such gates were already common in medieval mills. As well as simply releasing the flow, the gate needed to control the rate and strength of the water – a rise of as little as a foot above normal level in the stream could put the mill out of action. When the river was in full spate, control was vital. Before the Industrial Revolution, sluice gates were usually raised in order to open them, allowing water to flow underneath to the wheel.

Water dropped from the leat to the wheel and a slight incline from the headrace on to the wheel increased the fall, especially for an undershot wheel. The wheel itself was usually built into an enclosed wheel-pit, partly in order to increase the fall by excavating deeper below the channel. A second advantage of the pit was the increased efficiency that resulted from confining the water within a channel so that it could not flow round the wheel. The lining of the wheel-pit was often built of timber, though others were built of stone and brick, and often a composite of materials. There was no simple progression from timber to masonry construction – timber was still used in the eighteenth century – but certainly stone and brick were more common in the nineteenth century.

The water returned to the mainstream along the tailrace (also called the 'waste sluice' in some medieval documents). This was another important part of the channel, for the water needed to flow out quickly and cleanly to the river without clogging the wheel or backing up to the mill. Clearing water away from the wheel was especially necessary with undershot wheels, and it was usual to create a fall of up to 9 inches in the race immediately beyond the wheel. The tailrace flowed at a more gentle gradient than the river, and this meant that, in order to bring the mill water back into the mainstream at the correct level, some races were of great length – as much as half a mile or occasionally more. Usually, this could be achieved with a simple open channel, but it was not unusual to build the race in a culvert; this was the case with a number of the wheels used in the iron industry of the Weald.[47]

As well as the millrace, bypass channels were essential. These were used to divert water when the wheel was not working and to carry the surplus water when it was, especially providing relief in times of spate. Their efficiency could affect the success of the mill. Poorly constructed channels could result in rushing water altering the gradient at which the water reached the wheel or even undermining the foundations.[48]

The Technology of Water Power

The millpond for the corn mill at Cromford, Derbyshire.

Old wooden sluice gates preserved alongside Belvedere Mill, Chalford, Gloucestershire.

The Technology of Water Power

The gate to release water from the pond to the wheel for the textile mill at Lea Bridge, Derbyshire, equipped with metal filter to catch debris.

Taken together, the construction of weirs, millraces, ponds and bypass channels could result in some elaborate engineering.

It is not clear when weirs first began to be used for watermills. The Romans had built causeways as river crossings, and it has been suggested that St Albans Abbey might have used one of these as a dam in the late eighth century. Weirs at the time of Domesday were rare: one of the pioneer studies of Domesday mills commented on the lack of mills on the 'three great highway rivers' (the Thames, Severn and Trent), these being rivers on which management of the flow of water would be needed for mill wheels to work satisfactorily. However, there are some eleventh-century references to the construction of fish-weirs and mills. Hugh, Earl of Chester, built what is the oldest surviving mill dam in Britain on the River Dee shortly after the Norman

Water cascades away from the overflow channel at Tregwynt woollen mill, Pembrokeshire, 1959. (Museum of English Rural Life)

The Technology of Water Power

The weir at Ruswarp, north Yorkshire, as it is today. It is built at an angle across the River Exe to take water to the mill a few yards downstream, on the left. This is the site chosen for small-scale hydro-electricity generation.

conquest, to supply the mills of Chester with sufficient water. Ruswarp Mill was built at the tidal limit of the River Esk in Yorkshire by 1102, and a weir would have been necessary for this. Fountains Abbey, Yorkshire, was diverting the River Skell to its mill not long after the abbey's foundation, in the mid-twelfth century. On the River Derwent in Derbyshire, the monks of Dale Abbey built a weir in 1278. By the fourteenth century, weirs, often then called 'headweirs', and millponds were established features of the landscape. Little survives from the medieval and early modern periods, as older weirs were rebuilt, and many more were added in subsequent centuries.[49]

The number of weirs increased markedly during the early modern period, as water power was applied to a greater range of industries. At the same time, supply of water from leats became the preferred method over mainstream wheels, leading to the need for more diversions and digging of millstreams. On the River Derwent, the first weirs at Makeney were probably built during the sixteenth century, while the first weirs at Masson and Hopping Hill were built by the early eighteenth century. The construction of the new textile mills later in the eighteenth century added many more weirs and led to the rebuilding of existing ones. As well as being more numerous, weirs also became larger. On British rivers, weirs were modest in scale compared with other locations in the world. There was one in the Harz region of Germany that was 60 feet high and 475 feet long, whereas in Britain they were rarely more than about 10 feet high or more than 50 feet long.[50]

The principles of construction of leats for mills were well established in Saxon times, and some extensive works were undertaken. The millrace that led from the Thames to the Abbey Mills at Abingdon in 960 extended for almost three-quarters of a mile. Monastic estates were more prepared than most to undertake the construction of long and expensive channels. At Rievaulx, the abbey straightened the River Rye, to allow room for the leat to flow alongside. A diversion 5 miles long was dug in the twelfth century from the River Kennet to feed Reading's Abbey Mill. This stream, the Holy Brook, left the river at Theale and served a couple of other mills along its route to Reading. At Gloucester a similarly major diversion from the river created the Mill Avon

The Technology of Water Power

to drive the abbey mill. It was costly to undertake such works, and represented one of the major expenses of building a mill. Not every mill owner had sufficient capital, so that many a small medieval mill was left with a very basic water supply. Further expense might be needed following the construction of the weir in order to maintain the riverbank, for new currents created by the weir might start erosion. During the later Middle Ages estates reduced their expenditure on dams, weirs and leats. When a windmill could be built for as little as £10–12 there was little incentive to invest in a watermill. Canterbury Cathedral Priory spent £54 on dams, sluices, ponds and a watercourse of 1,887 yards in 1317, a sum that was unusual for its time. Revival in demand for water power meant that by the sixteenth century the mill that was served by a leat was becoming the most common arrangement, and during the centuries that followed there was renewed construction of complex water channels to serve mills.[51]

The dams and weirs built for mills were a potent cause of dispute over riparian rights with other river users, such as boat owners, fishermen and, in more recent times, water supply utilities. The more water that mill owners tried to draw off, the bigger the weir became and the greater the probability of a dispute. First amongst the opposing interests were the boatmen, whose clear passage along the river was being thwarted by the construction of weirs. Dale Abbey's weir of 1278 prevented boats from reaching Derby from the Trent and construction of a mill weir on the River Exe in 1313 effectively prevented navigation on the river. Regulation of rivers and mills was needed, especially, it was felt, to protect the navigation and trade of the nation. In the early thirteenth century the Lord High Admiral was given responsibility for English rivers, including regulation of mills. In 1422 the role passed to the Commissioners of Sewers.

Several laws were passed to control the building of weirs so as to protect navigation, and orders were made from time to time for the demolition of weirs. An Act of 1352 declared that all weirs built since the time of Edward I (1272–1307) should be destroyed. Its ineffectiveness was demonstrated by the fact that its terms were repeated in another Act of 1399. The weir on the Dee at Chester was one of many subject to orders issued on several occasions for demolition

Water pours over the high weir at Quarry Bank Mill, Cheshire. It was enlarged by Samuel Greg from 1801 as he developed the site for cotton-spinning.

or modification. Despite these orders; the weir is still there. Many sites for mills had been established when their rivers were not yet used extensively for trade, before interests in navigation had built up. When it came to it, mill owners had a ready ground for defence of their interests.[52]

The networks of channels were so complex in some places, especially in towns, that dispute was almost inevitable. At Newark it became difficult to distinguish the River Trent from its tributary the Devon, and from the artificial cuts to the mills. Medieval lawsuits tried to rein the millers in, for they had effectively captured the Trent. In 1558 the borough of Newark took the local millers to court because the diversion of water to the mills had left very little flowing in the river through the town.[53]

The boatmen had a real grievance: a weir was undoubtedly a hazard for boats. Until the introduction of the pound lock, the type that is now almost universal on British waterways, the most common means of taking a boat through the weir was a flash lock. An opening was cut into the weir, protected by boards. When they were removed, water would flow through the gap, raising the level downstream and lowering it upstream to allow boats through. Going downstream, boats would shoot the weir, as a canoeist might shoot rapids; coming upstream it was usually necessary to winch the boat. An alternative to the flash lock was the installation of gates, or half locks, that could be lifted clear of the river, or pulled sideways to create the opening for the boat. None of these was particularly satisfactory to either side. Mill owners resented the amount of water that was wasted, as far as they were concerned, every time a boat came along. Bargemen would complain of the niggardly attitude of the millers, making flash locks as inconvenient as possible for them, not allowing sufficient water through to float the boats, and then quite often charging to help them through. In the eighteenth century there were many complaints from Thames bargemen about the flash locks constructed mainly for the convenience of the mill, causing the larger barges then being introduced to catch on the cill of the lock.

Respite came only with the introduction of the pound lock. With this a separate channel was usually constructed for the boats to bypass the weirs and negotiate the changes in level on the river. This was an invention of some antiquity, but the first was not constructed in Britain until the sixteenth century, on the Exeter Canal. Although the use of pound locks spread from the seventeenth century onwards, flash locks remained common well into the nineteenth, even on such major rivers as the Thames. A few still remain.[54]

Management of fish was important in medieval times. Some of the weirs built were to supply fish-ponds. Fishermen were unsympathetic to the damming of rivers for mills, arguing that this impeded the free flow of fish. Sometimes a serious dispute arose, as in 1581, when armed men were hired by William Bowes of Streatlam Castle to break down the mill weir at Howsham, Yorkshire, to free the running for fish. Farmers also tended to be less well disposed towards mills, especially when they wanted water to improve pastures. Abraham Rees, in his *Cyclopaedia* of 1819, included a comment about farmers being irritated by the obstruction that mill works caused to the watering of land. Watermills 'should consequently be diminished in number … and those of the tide and wind kinds substituted in their place'. John Farey, who was probably the author of that article, further argued that the damage to fields caused by overflow from millstreams was significant: 'the annual value of damage done to adjoining Lands much exceeds the gross rental of the Mills.' The damage was greatest, he thought, in southern counties, where the leats in low-lying country were longer. He also included in his article in the *Cyclopaedia* the suggestion that dams and weirs supplying watermills should be reduced in number and replaced by longer cuts alongside rivers to serve them.[55]

Improvements made to the drainage of agricultural land could also work to the mill's disadvantage. Better-drained fields did not always deposit their outflow water into the stream conveniently for the mill. Large-scale drainage works to bring new land into cultivation were a particular problem. The Lake of Kinnordy in Scotland was drained in 1740, which resulted in the Dean river diminishing from a 'considerable stream' to a flow that was 'scarcely sufficient to turn a mill'.[56]

The effect of one mill's dams on other mills caused as much if not more trouble than problems with other river users. From the earliest written records about mills, such problems start to appear. In the eleventh and twelfth centuries Abingdon Abbey was involved in a number of disputes with its neighbours, both upstream and down, relating to the effect of each other's weirs on the water flow to their mills.[57] Problems could arise as soon as a miller made any alteration to improve his supply. If he raised his weir to build up more water behind his wheel, the miller downstream and the one upstream could be affected. The one downstream could find his supply reduced because the water was not flowing quickly enough away from the upper mill. Upstream, the water might be backed up behind the enlarged weir so far as to flood the tailrace, preventing the wheel from turning. A full reservoir of water backing up into the tailrace upstream is known to have affected Joseph Rishworth's woollen mill at Elland, Yorkshire, in the early 1830s.[58]

As rivers filled with mills, the problems increased: fitting a new mill on to a busy stream could quite often cause trouble. A new fulling mill on the River Wandle was ordered to be pulled down in 1693 after the operator of a mill downstream grinding brazil wood for dyes complained that his water supply had been cut.[59] Richard Arkwright built a cotton mill at Lumford on the River Wye just upstream of Bakewell in 1777. It was worked by an undershot waterwheel supplied from a reservoir of 5 acres. He incurred the wrath of the Duke of Rutland, who contended that his corn mill at Bakewell had been starved of water. A long dispute ended in 1786. Arkwright was forced to make restitution to the duke, but he still had his Lumford mill at work.[60]

No formal legislation was laid down about the rights of millers; instead, engineers and millwrights had to work to a few general principles. In the eighteenth century, the engineer John Smeaton set these out as follows: 'dams must be so built that no one shall pen the water into the wheel race of the mill next above, when the river is in its ordinary summer state.' Basically, every miller had the right to use the water passing through his land, provided he did not raise the level upstream so as to block the mill above, or divert water on the outflow so as to starve the next mill downstream.

According to convention, responsibility rested with the miller downstream to protect his own supply of water, which meant that he had to build a larger reservoir if necessary. That did not prevent disputes, which were frequent. Samuel Greg, who owned Quarry Bank cotton mill at Styal, received a complaint from a miller lower down the river in 1815 that the expansion of Quarry Bank had adversely affected the lower mill. The matter was resolved between the two, and this was the most common way to proceed if the courts were not to be involved. The mill owners, often together with their landlords, would come to an agreement to limit the effect of one mill upon its neighbours. The owners of Ebley and Dudbridge mills, in Stroud, agreed in 1685 to restrict the amount by which the height of the weirs might be raised. Occasionally, disputes did go to court. The dramatic possibilities of such cases were exploited by George Eliot in her great novel *The Mill on the Floss* – it was a dispute with a neighbour diverting water upstream that caused the downfall of Mr Tulliver.[61]

MILL GEARING

The observer of a water-powered mill sees an apparently complicated set of gear wheels and shafts deriving from the waterwheel. These perform two functions. One is to transmit the drive from the waterwheel to the millstones or other machinery, on the way converting the drive from the vertical plane to the horizontal as needed. The second is to transmit the power at the appropriate speed for the machinery, because it is rare for the waterwheel and machinery to revolve at precisely the same speed. In corn-milling, since the eighteenth century at least, it has been usual for millstones to be run quite quickly, at about 120rpm, while the vertical waterwheel turns quite slowly. The medieval mill turned the stones more slowly, at a speed somewhat above 50rpm, but still faster than that of the waterwheel. The intermediate transmission in both medieval and

The Technology of Water Power

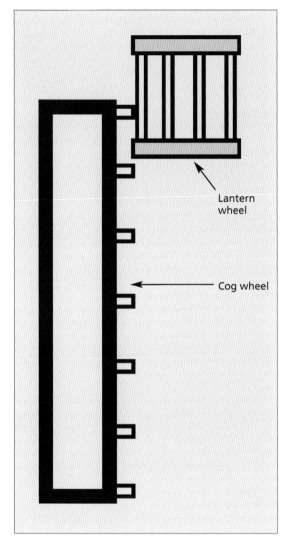

Drive from cogwheel to lantern pinion, or trundle wheel, that was the standard in medieval mills, similar in many respects to the Roman mill depicted by Vitruvius.

turned at a much higher speed than was usual in a later wheel, necessitating step-down gearing. Exactly when it became usual for gearing to step the speed up from wheel to stones is not certain.[62]

The simplest form of power transmission was that of the horizontal wheel, from which a shaft could drive from the axle direct to millstones also turning in the horizontal plane without any intermediate gearing at all. The building had to straddle the stream, as did many a mill driven by vertical wheel, but this did preclude the use of a horizontal wheel if the mill could not be built across the stream. Direct drive to the millstones required little maintenance, which suited the small mill, as repairs could usually be handled by the farmer or miller.

The vertical wheel required gearing that transferred power to the horizontal plane for the millstones. Although this need not be elaborate, intermediate drives, with their toothed or cogged gear wheels, were likely to be more complicated than the miller might wish to handle, and the skills of the carpenter or millwright would be required. For corn mills, trip hammers and pumping, simple arrangements of shafts, wheels and cams worked. As water power was applied to industries such as textiles, with greater demands for power and more machines to be operated, gearing became more complex.[63]

The drive from the vertical wheel in a watermill of medieval and early modern times was via the axle shaft that linked waterwheel and gear wheel. It was long – 12 or 15 feet – and about 20 inches in diameter, formed usually from a single oak trunk. Supplies of timber to meet this specification were always limited, and in consequence the wheelshaft was an expensive item.

As well as describing the vertical waterwheel, Vitruvius also elaborated upon the means of power take-off from it: 'At one end of the axle a toothed drum is fixed. This is placed vertically on its edge and turns with the wheel. Adjoining this larger wheel there is a second toothed wheel placed horizontally by which it is gripped. Thus the teeth of the drum which is on the axle, by driving the teeth of the horizontal drum, cause the grindstones to revolve.'[64]

Essentially, this was the type of drive used

modern mills, therefore, steps up the speed from wheel to stone. This was not always the case. The watermill described by Vitruvius, the Roman writer, clearly has transmission that steps the speed down from the waterwheel to the millstones. Some modern students of his text have treated this as a mistake, and assumed that he meant to describe gears to drive the stones at a faster rate, but the general accuracy and detail of his work leads others to accept what he wrote. Some archaeological evidence also suggests that waterwheels of the Roman period

The Technology of Water Power

throughout the medieval centuries. Constructed of wood, the vertical toothed wheel (the 'cogwheel') was built up rather like a waggon wheel, with felloes linked together to make up the circumference. The second, horizontal toothed wheel, with which the vertical wheel engaged, was of similar construction. One variant, which seems to have been known in Roman times and certainly became common in the medieval period, was the lantern wheel. It was double-skinned, the teeth making up the supports between the upper and lower panels (*see page 45*).

In the Middle Ages the lantern gear was known as a 'trundle', and was usually of small diameter. It was mounted on a large raised beam (the 'bridge tree') so that the vertical cogwheel engaged with it at the 12 o'clock point. At the centre of the trundle was the iron spindle that passed up through the centre holes in the stones to engage with the 'rynd', the iron bed on which the upper millstone was settled, and thus to turn the stone. The wheels were of very solid construction, and could cost nearly as much as the

Nether Alderley Mill, Cheshire, was built in the sixteenth century. It had two, possibly three, wheels. The present wheels and gearing are of more modern installation.

waterwheel. A waterwheel made for the mill at Castleford, Yorkshire, in the fourteenth century, cost 10s, while the cogwheel cost 6s 8d. Trundles were tied with iron bands, making them heavy: a weight of 48 pounds was mentioned in a fifteenth-century inventory at Winchester.[65]

This type of gearing could drive no more than one pair of millstones. To increase the capacity of the mill it was necessary to install an additional wheel. In the Middle Ages there were a number of examples of a complete additional mill being built, with wheel, gearing and stones. Quite often, the new mill was accommodated by extending the existing building, so that the whole looked like one, although medieval records would always refer to them as separate mills.[66] After the Black Death, such extensions became uncommon, but, as demand grew during the late sixteenth or seventeenth centuries, there is clear evidence that many mills were rebuilt or extended to accommodate the additional wheel. Estate surveys of this time also refer to the mills in terms such as 'two corn mills under one roof'. The most common arrangement was for the wheels to be side by side, or on opposite sides of the building. The medieval mill at Abbotsbury, Dorset, had two wheels alongside each other at some stage in its development. Occasionally, wheels were installed in other ways. At Nether Alderley Mill, Cheshire, built in 1591, the overshot waterwheels were placed in line with each other in order to fit into the site. The first was at a

A close-up of the hand mill shown on page 13, showing the lantern drive that was still used in this type of mill in the eighteenth century. (Museum of English Rural Life)

The Technology of Water Power

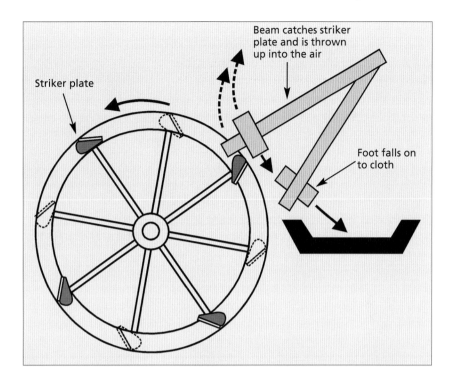

A simplified diagram showing the working of a cam wheel to operate fulling stocks.

higher level than the second so that water fed from one to the other. There may have been a third wheel alongside the second, although that has long gone.[67] As well as the additional power, there could be other advantages in having more than one wheel, for the mill might then be used for more than one purpose. A number of medieval mills combined corn-grinding and fulling in the one building, each powered by a separate wheel.

Some of the early industrial uses of water power during the later Middle Ages were driven by similar types of gearing. Crushing bark and gunpowder mills, for example, worked on the same rotary motion as flour-milling. Some used edge-runner stones instead of horizontal stones, but the drive was the same. Fulling mills and hammer mills in iron-making, however, involved turning the rotary motion of the waterwheel into a reciprocating action. This was achieved by using the cam, which is a peg projecting from an axle. As the axle turns it can trip a hammer or other tool into action. This axle, the camshaft, could be driven directly from the waterwheel. The cam was an invention of classical times, but the medieval fulling mill gave it a bigger economic role.

At the end of the Middle Ages, the offset crankshaft, another old invention, known in ancient China, was adapted to drive bellows from waterwheels.

MILLWRIGHTS

The millwright was the craftsman responsible for installing the waterwheel and the mill gearing. The Irish language had a word for the millwright as a specialist as early as the seventh century, but most documents of the medieval period refer to the builders of mills as 'carpenters'.[68] That was probably an accurate description of how the trades operated: the carpenter added the building and maintenance of mills to his skills. While most remained in general trade, there were sufficient mills by the later Middle Ages for some to develop more specialist skills. The word 'wheelwright' emerged in the later fourteenth century. By that time the numbers of fulling mills and hammer mills for iron working were increasing to add to the corn mills to build and maintain. From this time the millwright became a more recognized trade. By the second half of the eighteenth century, when trade directories started to be published, numbers of millwrights were significant. There were twelve millwrights listed in the directory of 1820 for Gloucestershire.[69]

The millwright's work was never tightly defined, and nor was the route into the business. In the woollen district of the West Country, many of the millwrights identified were from the families of clothiers who owned the mills; elsewhere, millwrights might have a background in wood-working trades.[70] The millwright worked mainly in the wood with which he built the wheels and gears and fulling stocks. But he also had to turn his hand to engineering, in the construction of dams and watercourses. Primarily, however, the millwright's trade was still regarded in the eighteenth century as a branch of carpentry. Sir William Fairbairn, who became one of the greatest of millwright-engineers of the nineteenth century, looking back on his apprenticeship, observed that the wheelwright of the eighteenth century was 'a kind of jack-of-all-trades, who could with equal facility work at a lathe, the anvil, or the carpenter's bench'.[71]

Millwrighting was a craft learned through apprenticeships and local insight rather than schooling and science. The wheels and mills constructed were similarly based not on calculation of the force of water and power required for the mill, but on intuitive knowledge of the river and the needs of the works in hand.

Most millwrights worked on a local scale, but by the end of the seventeenth century a distinct hierarchy was emerging, with men such as George Sorocold, 'expert in making mill-work', as Daniel Defoe described him, also becoming noted hydraulic engineers. Sorocold won contracts in many places around the country, and this trend in the eighteenth century saw the rise of the consulting engineers, such as John Smeaton, whose technical and scientific skills won them work on a range of projects.

MILL BUILDINGS

Clearly, milling was a capital-intensive process. Investment went primarily into building the wheel and its gearing, the dam and watercourses. The building that housed the processes also needed some special consideration, for it had to withstand both wetness and vibration. These requirements on their own were enough to make the medieval mill usually a distinctive building of some substance, made prominent by its riverside site and wheel.[72]

In other respects, the medieval and early-modern mill did not always stand out, being built in the local vernacular, in common with most of its neighbours. It was not usually particularly large, unlike its grand multi-storied successors of the eighteenth and nineteenth centuries, mainly because it needed less storage. Most of the trade was for the manorial

The watermill with its overshot wheel shown in the fourteenth-century Luttrell psalter. (Museum of English Rural Life)

tenants who brought their corn and took away their flour, so the miller did not need to keep the product on his premises for very long. The sixteenth-century mill at Nether Alderley, Cheshire, is typical of the scale of building (*see page 46*).

Where it was available, stone seems to have been a not uncommon building material for medieval mills; the mill illustrated in the Luttrell Psalter was evidently built in this way. Away from the districts of stone building, most mills were likely to be of wattle and daub construction, with perhaps some timber cladding alongside the water channels to prevent erosion. Thatch was the most likely roofing material, referred to in many a survey of medieval and early modern centuries, although at that time 'thack' might also mean a roof of stone slabs.

Although the sites of many mills recorded in Domesday Book are known, the history of rebuilding many times over means that nothing remains of the mill buildings. Very little can be identified of later medieval watermills, not even the stone-built ones. Fountains Abbey mill in Yorkshire and the Old Malthouse at Abbotsbury, Dorset, are two survivors. At Fountains, the surviving building, although not the original on the site, is of largely thirteenth-century construction. The mill was used with water power by the Fountains Abbey estate into the twentieth century to work a sawmill, with a turbine installed.[73] Excavations at the Old Malthouse at Abbotsbury revealed the building to have been a mill built in the fourteenth century for the neighbouring abbey. It might not have been the first mill on this site, and it was certainly altered later, with part of the building converted into a dwelling, although its use as a mill continued until the eighteenth century.[74]

Damage from the stream and from the vibration of the wheel and millstones was a continual problem. Repairs were regularly needed, as many manorial accounts record, and this is something that contributed to the paucity of surviving mill buildings from that time. There have been periods of major rebuilding, first, from the sixteenth to the early seventeenth centuries, when larger buildings replaced many medieval mills. Almost all followed vernacular building traditions, using the materials of the locality. Generally they were likely to be built of

The medieval mill at Abbotsbury, Dorset, known now as the Old Malthouse. Archaeological work suggests that at one time the mill housed two parallel waterwheels each driving a set of stones. (Alan Stoyel)

stronger, longer-lasting materials. Most were probably still being built of timber, of such quality that a number of timber-framed mill buildings survive from this period.

Inside, the layout of the water corn mill was quite simple. The principle of working by gravity was established early on: grain was stored and cleaned on upper floors, moved down to the grinding floor and on to the flour store. The sack hoist, often with its projecting lucam, became a characteristic feature of the corn mill. The space taken up by the wheel and gears might limit ground-floor storage for flour; otherwise, this layout offered a smooth flow of operation that changed little before the Industrial Revolution. The size of a corn mill was determined as much by its need for storage and ancillary operations, such as cleaning the grain, as by its gears and stones.

TIDE MILLS

Mills that use the power of the tide to turn the wheels are of some antiquity. The earliest known example in

Rossett Mill, near Wrexham, was built in 1661. It is of timber-framed construction with brick infill, and some stone to support the undershot wheel and strengthen some parts of the building by the stream. When photographed in the 1950s it was still a working corn mill; it is now preserved as part of a residence. (Museum of English Rural Life)

the British Isles (and indeed Europe) was discovered during archaeological excavations on Strangford Lough, Northern Ireland. The remains found there were of a tide mill with a horizontal waterwheel, dated to c.619–621AD. It was replaced, it seems, by a second mill built on the site c.789. These mills were built for Nendrum monastery. It was not the only tide mill in Ireland from this time – some others have been dated to the seventh century – but evidence from elsewhere is less forthcoming. An Anglo-Saxon charter of 949 referring to a mill near Reculver in Kent implies that this was a tide mill. There is a reference in Domesday Book to 'a mill which causes disaster to vessels by the great disturbance of the sea and so causes the greatest damage to the king and his men' at the entrance to the port of Dover; some interpret this as meaning a tide mill. There is much clearer evidence of tide mills later in the medieval period, most of them belonging to monastic estates, such as Westminster Abbey's mill and one at Wapping belonging to Aldgate Priory. At least thirty-seven had been built in England by 1300. In Devon and Cornwall there were three, increasing to five in the fourteenth century and nine in the sixteenth. By the sixteenth century there were more in evidence on tributaries of the Thames at Rotherhithe and Deptford, and at what is now known as Three Mills at Bromley-by-Bow.[75]

Tide mills need a gentle, steady flow of water. A mill built directly on the coast would be broken up by the waves. The mills could therefore be built only off low-lying coasts, with plentiful shallow creeks, and this is exemplified by the half-dozen tide mills of Essex. Thorrington Mill and St Osyth Mill, for example, were both sited on creeks off the River Colne, while Stambridge Mill was off the River Roach. Tide mills were a feature of the south coast of England from Essex as far west as Devon and Cornwall, on tributaries of the Thames and in Pembrokeshire. A mill at Bidston in Cheshire was one of only a few further north. As well as the mills at Nendrum, about twenty early medieval tide mills have been identified around the coast of Ireland. A number were built in Northern Ireland in

The Technology of Water Power

Thorrington tide mill, Essex, of weather-boarded construction, stands on the Alresford Creek, off the River Colne.

The partially enclosed undershot wheel at Thorrington.

The Technology of Water Power

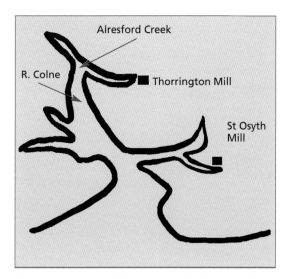

Map showing the siting of Thorrington and St Osyth tide mills in Essex on creeks off the River Colne.

modern times. The first recorded was in 1744 at Castleward. In 1824 a tide mill was mentioned at Saltwater Bridge, Finnebogue, while two further mills were built on Strangford Lough, at Ardchin and Strangford Lower.[76]

Tide mills were almost all grain mills. Among the few exceptions was a medieval tide mill in Suffolk used as a fulling mill. There was a mill for boring cannons at Hackney Marsh near London, one of the two mills at Bidston was an iron-slitting mill, and one of the mills at Bromley-by-Bow was apparently used briefly for gunpowder work in the late sixteenth century.[77]

The Three Mills at Bromley-by-Bow, east of London, were among those recorded in Domesday Book, and have had a varied career. Despite the name, there have only been two mills on the site since the sixteenth century, the Clock Mill and the House Mill, tide mills on the Three Mills Wall River, a channel off the River Lea. The mills were bought in 1734 by Peter Lefevre, a distiller who used them mainly for grinding malt, some of which was produced in maltings that were added to the site. House Mill was rebuilt in 1776 with four wheels driving eight pairs of stones. It was still in the ownership of distillers in 1886 – the business was now Nicholsons – when the economy of water power for the malt-grinding was enough to justify another rebuilding. A new cast-iron wheel was installed in the House Mill, two of the other wheels had curved floats added, and the stones were given new drives and gearing. This was one of the last major reconstructions of a water-powered grain mill.[78]

The Three Mills, Bromley-by-Bow, east of London. Both of these buildings were tide mills, but the one on the right, Clock Mill, was used later as a malting for the distillery.

Eling tide mill on a creek off Southampton Water. The water flows into the reservoir through a gate to the left of the picture, coming out through the mill wheel enclosed under the building.

The operation of these mills was, of course, determined by the tide. Most, certainly all of those in Britain, were worked by the ebbing tide. Water from the incoming tide was captured in pounds, which had gates that worked on a principle similar to that of the lock gate. The pressure of the tide coming in pushed the gate open. When the tide turned, the pressure worked the opposite way and forced the gate shut. The water could then be channelled out to work the wheel. Some of the reservoirs were extensive, a result of the low-lying sites for these mills: at Birdham in Sussex, the reservoir covered 13 acres, and at Walton-on-the-Naze in Essex, it extended to 30 acres. Waterwheels were usually undershot, as the head was often low. Eling mill, near Southampton, installed the improved Poncelet type of undershot wheel in the nineteenth century. Some were in better circumstances and were able to install breast-shot wheels in the eighteenth and nineteenth centuries. Fingringhoe in Essex had a breast-shot wheel that was 16 feet diameter by 8 feet wide, which drove three pairs of stones. Tide mills tended to be modest in scale, most working two pairs of stones, although some, such as those at Woodbridge, Suffolk, and Pembroke, operated four pairs.

A tide mill had the advantage of a guaranteed supply of water subject neither to drought nor freezing. Tidal surges or storms could damage the dams, although their sites on inshore creeks could mitigate that. The big disadvantage of the tide mill was that working time was limited to a few hours before and after low water. This also meant that the miller might be at work in the early hours of the morning as much as mid-afternoon. Some attempts were made to alleviate this by building bigger pounds, so that the mill might work a little longer as the tide ebbed, but the most that might be expected in a session would be about eight hours. Most mills could manage no more than five or six hours with each tide, and neap tides might make matters worse. Another way to enable work to continue regardless of the tide was to have a windmill close by, as at the Essex mills of Thorrington and Walton. By the nineteenth century there were means of employing both the ebb and flow of the tide, enabling the mill to work round the clock. However, this involved having another gate to the reservoir or a second waterwheel, so for most mill owners it was not worth the effort, especially when competition from steam power was growing.[79]

5 Flour-Milling Before the Industrial Revolution

Most of the early medieval watermills were built to supply the needs of their immediate locality, the manor or the monastic estate. This was not exclusively the case: there was some trade in flour, but it was on a modest scale. The exceptions were the towns, where the population needed supplies in greater quantities, but even they were supplied from relatively local mills. The introduction of windmills helped, by enabling additional sites to be used, especially where rivers through towns had a low fall. A ring of windmills on the outskirts and close suburbs became a characteristic feature of many towns, and this lasted well into the eighteenth and nineteenth centuries. London, by far the biggest city in Britain, was able to draw most of its supplies from near by. Apart from the scale of demand and population, there was a simple fact behind this: grain has better keeping qualities than flour, so in principle it was always better to make the grain travel further than the flour. Flour mills would, therefore, be sited as close to the market as possible.

The restrictions placed upon a miller engaging in trade meant that it was the manorial estate or its tenants that sold grain or flour to the bakers of the town. By the late fourteenth and fifteenth centuries this was changing. London and some of the other large towns were needing to draw supplies of flour from further afield. During the centuries after the Black Death, as the enforcement of manorial rights over mills weakened, a new hierarchy emerged in the milling trade. Most mills had a local trade and continued to operate on a toll system, whether the

Ham Mill, Newbury, one of the many commercial flour mills on the River Kennet with trade to London. This photograph was taken in the first decade of the twentieth century. (Museum of English Rural Life)

feudal suit of mill was still being maintained or not. Many country mills were still trading in this fashion in the nineteenth century. Worsborough Mill in Yorkshire was one example, only converting from toll to merchant milling in the early nineteenth century.[80]

Some millers by the sixteenth century were becoming merchant millers, who bought their stock of grain to mill and built up business on a larger scale supplying flour to London and other major towns. At that time they were acting illegally because they were still forbidden to deal in grain and meal, but there was rarely any enforcement. Every so often in the sixteenth century there would be an outcry, usually after a poor harvest pushed up the price of bread; government said it would act, but once the immediate crisis had subsided things carried on. The merchant millers continued to expand their trade, to prosper, and to make their mills bigger, with space for the grain and flour they now had to store.

The need for increased supplies of food lay behind London's experiments with floating mills (*see* pages 33–4). There were two attempts in the sixteenth century to establish floating mills in London and another two at the end of the eighteenth. One entrepreneur, John Cooke, had two floating mills in operation in 1519, and had added another two by 1523. Thereafter, nothing is heard of them; presumably they ceased operation after a short time. That was certainly the fate of the next attempt with a floating mill, at Queenshithe at the end of the sixteenth century, which also lasted just a few years. In 1793 a floating mill started operation, but the authorities decided after a few months that it obstructed navigation and fishing. In 1804, Benjamin Hawkins obtained permission to site a mill along Bankside between Somerset House and Blackfriars Bridge. This mill operated until 1811, when John Allen, who had succeeded Hawkins as proprietor, went bankrupt.[81]

The growth of the city of London brought about a distinct pattern of trade. In the first place, mills continued to operate in areas that by the twentieth century had come to be regarded as part of the central urban mass, not even inner suburbs. Most of the city's flour was brought in from the rural home counties.

Daniel Defoe in *The Complete English Tradesman* described the trade as it had developed by the early eighteenth century, when flour was being brought into London from a radius of 40 miles. He described Guildford as a major market for grain, with wheat going from the town to the mills on the River Wey, and thence to London. Upriver from London he singled out Marlow as an important town in the trade in grain and flour, drawing supplies from the High Wycombe area for grinding at the Thames-side mills. The mills along the Essex waterways, the Medway in Kent and along the Thames and its navigable tributaries were also becoming major sources of supply for the capital. The limits to London's catchment area were determined principally by transport: places such as Newbury and Abingdon were beginning to be drawn into the orbit, as were mills further into East

A small country mill, Hele, near Ilfracombe, at the beginning of the twentieth century. (Museum of English Rural Life)

Anglia, on such rivers as the Waveney and the Stour. Some of the towns involved in this trade, such as Chelmsford, Reading and Marlow, were local milling centres, where millers were part of local business networks alongside farmers, grain dealers, maltsters and owners of barges and wagons. Some returns of production during the Napoleonic Wars reported a mill in Reading with capacity to grind 25,480 sacks a year; four mills at Newbury could produce a total of 14,560 sacks.[82]

There were clear watersheds. The limitations of water transport kept the mills of Wiltshire out of the London trade, while only mills on the lower reaches of East Anglian rivers had a share. On the Thames, Abingdon was the limit beyond which trade developed during the eighteenth century in the opposite direction to supply growing demand from the Midlands towns. In the 1790s, about a third of the output of Wallingford's corn mills went to Birmingham. Mills on navigable rivers in southern counties, Sussex and Hampshire, for example, might develop a trade in flour to the West Country, but that was limited by the fact that the west of England was itself also good for water power. There were many mills in Devon able to supply Exeter. In the north-west, Liverpool and Manchester drew supplies from the watermills of Cheshire and south-west Lancashire. Mills at such places as Chester, Warrington and Lymm were able to expand on the strength of this trade.[83]

Still, the majority of mills did not share in the urban trade. Theirs was a much more local business. Small mills could readily serve three or four villages within a radius of 10 miles or so, while market towns all had their mills within close proximity. These mills supplying their local markets also drew their supplies of grain from within about the same radius, at least until industrial times, when millers in more pastoral areas might have to look further afield for their grain. A large number of mills operated almost entirely within their locality. About 70 per cent of the deliveries of grain to Fairbourne watermill in Kent in the mid-eighteenth century came from within a 3-mile radius.[84]

The smallest of mills in the Shetland Islands, working on little more than domestic scale. When this photograph was taken, in the 1940s, it had been out of use for some time. (Museum of English Rural Life)

The small, rural mills were not necessarily very profitable. In manorial records their values in the early seventeenth century are often little changed from those in the fifteenth.[85] The millers were as likely as not to have other occupations besides milling, often farming or malting, but many another trade was carried on alongside the milling.

By the beginning of the eighteenth century the nature of the grain-milling trade was changing. The custom of farmers bringing their grain to the mill to be ground while they waited, which continued by and large well beyond the medieval period, was breaking down. Farmers were more likely to leave grain to be milled at the miller's convenience. The miller in consequence had to have greater storage capacity. It was a natural progression for him to start trading in grain and flour, especially if he was involved in trade to the large towns.

6   Early Industrial Uses of Water Power

One of the questions twentieth-century historians have raised is how much medieval economy and society looked to mechanical power. Professor Lynn White in an article published in 1940 argued that there was a revolution in the use of power during these centuries, while Eleanor Carus-Wilson in 1941 made the case for an 'industrial revolution' in England in the thirteenth century, based on water power.[86]

One of the strongest points in support of their arguments is the way in which the use of water power expanded beyond flour-milling to a wide range of other industries and mining. Almost anything that involved crushing, grinding, hammering or lifting of water it seems was driven by a waterwheel somewhere or other. It was a perfectly logical step to apply the waterwheel to other purposes besides grinding corn and that step was taken quite early on. The Roman writer Ausonius refers to a sawmill for cutting stone driven by waterwheel on a tributary of the River Moselle in France in the late fourth century. In common with many of the early references, its interpretation is not perfectly clear, but most authorities now accept it as genuine. It has been suggested that the Roman mill site at Ickham in Kent might have used water power for metalwork, but the evidence is far from conclusive, and the same can be said of the mill at Muncaster in Cumbria, and other places put forward as pre-medieval users of industrial water power. Evidence for industrial mills throughout Europe before the late ninth century is lacking, and before the twelfth century they appear to be isolated instances at best, with considerable doubt as to whether they did indeed use water power.

Certainly in Britain there was almost no industrial use for water power before about 1200. Most of the pioneering developments took place in continental Europe some time before they were taken up in Britain. In particular, the alpine regions of northern Italy, southern Germany, Switzerland and south-eastern France, where water power was plentiful, and parts of northern France, were in the forefront of the growth of industrial water power from the twelfth century onwards. In Britain, demand for most industrial products was not sufficient to justify investment in capital-intensive applications. Because of this weak demand, the return on a mill used for industry was likely to be less than that of a corn mill. The lord of the manor, therefore, would prefer a lease to a corn miller at a higher rent than he would get from a mill used for other purposes. Where demand for grain-milling capacity was strongest, in East Anglia, for example, there were few industrial mills in medieval times.

Significant change came to England from the late twelfth century onwards, and this was described by Eleanor Carus-Wilson as an 'industrial revolution'. In common with the more familiar Industrial Revolution of the eighteenth and nineteenth centuries, this one was based on textile manufacture, and its motive power was water. Only one process in the woollen industry, fulling, was the subject of mechanization, but the water-powered mills soon became an established feature. This was not the only industrial application of water power to be taken up in medieval Britain. One of the earliest, almost contemporaneous with the fulling mill, was the bark-crushing mill for leather-making. The first of these has been noted from about 1165. Only a handful were built by the end of the Middle Ages, although they became more important later. In some localities they were significant, one example being

57

the Kennet Valley. There were mills at Newbury, Shaw and Speenhamland in the early sixteenth century, crushing bark for tanning and for parchment-making. Maurice Smart, one of the occupiers of tanning mills (as the crushing mills were alternatively known), was the wealthiest resident of Shaw in 1522. Iron working, mining, paper-making and gunpowder were among the other applications of water power before the Industrial Revolution.[87]

Two main phases in the expansion of industrial water power after its introduction during the twelfth and thirteenth centuries have been identified. The first was concentrated in the second half of the fourteenth century. The decline of population after the Black Death increased the cost of labour, creating an incentive to use water power, while lower demand for corn mills provided an opportunity in that there were mill sites available for conversion to industrial use. In 1300, 6 per cent of the total number of mills in England and Wales were employed for industrial processes. Growth later in the century took the proportion of industrial mills up to about 12 per cent by 1450. Economic recession during the middle decades of the fifteenth century resulted in a slowing of the expansion of industrial water power in England. Nevertheless, in the fifteenth century, although they might not have challenged the dominance of flour-milling, watermills for industry were noticeable.

The second period of great expansion started about 1490 with renewed economic growth. By 1540 the proportion of mills employed for industry is estimated as 23 per cent.[88] From this time onwards, the expansion of water-powered industry continued. It was stimulated by demand from large towns, while a new generation of entrepreneurial landowners, many of whom had bought former monastic property, was keen to exploit their estates. This can be seen in some local examples. In the Tillingbourne Valley in Surrey there were nine water corn mills or fulling mills at the end of the fifteenth century. During the sixteenth and seventeenth centuries, a further twelve were added, most of them for industrial use – gunpowder, brass and wire mills, hammer mills, paper mills and sawmills. Closer to the demand of London, the River Wandle experienced similar growth. A new cut along the river was made at Merton, among many works undertaken to enable additional mills to be built, and existing mills to be rebuilt and enlarged. Growth continued into the eighteenth and early nineteenth centuries, with the addition of snuff mills and calico-printing works, and the Wandle was frequently remarked upon for the concentration of mills along its course.[89]

THE FULLING MILL

The industrial development that prompted Carus-Wilson to put forward the concept of a medieval industrial revolution was the mechanization of fulling. Fulling was a process in the manufacture of woollen textiles, in which woven cloth was subjected to a pounding to remove grease and impurities, and to help stretch and bind the fibres together. It required water to rinse the cloth and an agent to aid the cleansing. The detergent most often used was fuller's earth, a type of clay that helped in the binding. Where it was not available, other substances, mainly urine or a lye from the ashes of bracken, were used. Above all, fulling required the application of pressure and pre-mechanization this meant either beating by hand or with wooden mallets, or, most often, trampling the cloth underfoot. (These processes of fulling and 'walking the cloth' led to such medieval surnames as Fuller and Walker.) It was a long process: even after the water-powered mills were introduced, it could last eight hours or more.

The work was turned into a mechanical activity with the introduction of the fulling mill, or fulling stocks, known in the West Country as 'tucking mills', and in some other places as 'walk mills'. Large hammers were worked by rotary action off a waterwheel. The hammers, set in pairs, were raised alternately to the horizontal position by cams, which tripped the handles. From this point they dropped through approximately 90 degrees on to the cloth. It was the first use of cams in this way to convert rotary motion into linear oscillating motion, and was a major step in the application of water power to industrial processes. Another similar tool in the

Fulling stocks at Hunt & Winterbotham's mill, Cam, Gloucestershire, in 1963, not long before they were dismantled and taken to Stroud Museum. (Museum of English Rural Life)

textile manufacture was the stamper, used mainly for the felting process. This was a heavy piece of timber lifted vertically by water power and allowed to fall under its own weight to pound the cloth. The same principle was followed shortly by the use of other stamping machines in the manufacture of paper, gunpowder and other products.

Introduced in Italy in the early eleventh century, or maybe even the late tenth century, water-powered fulling soon spread across Europe. The first fulling mills in Britain known with certainty were at houses of the Knights Templar at Barton on Windrush, Gloucestershire, and Temple Newsam, Yorkshire. References to both of these first examples are found in records from 1185; there might have been one or two earlier examples, but the records are not clear. The international connections of the order of Knights Templar probably helped bring the technology to England. In general, the leading estates in the adoption of water fulling mills during the thirteenth century were monastic and ecclesiastical. By the end of Edward II's reign (1327) there are known to have been 124 of these mills.

Whether that growth constitutes an industrial revolution, as Carus-Wilson argued, has been challenged. Impressive though the thirteenth-century growth in numbers is, it represented just a small part of the woollen industry, and many historians believe that it falls short of being 'revolutionary' in its impact. The substantial investment needed in the heavy fulling stocks for what was only a small part of the whole process of cloth manufacture was probably not a viable proposition, except for the large estates, most of them monastic. It was not until the second half of the fourteenth century, after the Black Death, when more significant change took place in the woollen industry, that the fulling mill began to come into its own. A shift from the export of raw wool to the export of manufactured cloth coincided with the development of a heavyweight broadcloth, which came to dominate the later medieval wool industry of England. The fulling mill and the water power that drove it played a major role in that development. The mechanical mill could run for much longer than the manual process, and was thus more useful for the heavier cloths. There was official encouragement as well, with the passing of an Act of Parliament in 1377 forbidding the export of cloth that had not been fulled.

These developments were enough to prompt a more rapid increase in the number of water-powered fulling mills from the later fourteenth century onwards. Operation of the mills began then to widen, to include larger numbers of independent clothiers. There were two fulling mills at Newbury in the fifteenth century, the Town and West Mills, operated by local merchants, who also leased mills in the surrounding countryside. John Bennet, who leased the fulling and bark-crushing mills at the town in 1436, was one of the wealthiest men in the locality.[90]

Another aspect of this medieval industrial revolution is the way in which water-powered fulling mills affected the location of the woollen industry. Carus-Wilson argued that water power took the industry out of the towns in which it was established, so that it became more based in the country, a move aided by opposition from vested interests in towns. Other scholars have found problems with this analysis. Good sites for water power were certainly not confined to towns, so, naturally, there were many rural mills. There were also examples of towns that did not welcome fulling mills – Bristol was one – but there were others, including Salisbury and Oxford, where water-power sites in and immediately around the town were turned over to fulling mills. At Exeter, the Guild of Weavers, Tuckers and Shearmen, founded in 1471, promoted the development of fulling mills along the Exe as it flowed through the town.[91]

What is clear is that there developed quite a specific geography of water-powered fulling. Medieval fulling mills were found mainly in a swathe of the country from the West Riding of Yorkshire down to Wiltshire. The largest number of medieval mills was in the south-west, from Gloucestershire to Cornwall: forty-six of the thirteenth-century mills were in these counties. Local concentrations of fulling built up along rivers where the supply of water was plentiful and the flow constant, and this had a strong influence on the subsequent development of the textile industry in this region. In Gloucestershire, the Cam, the Frome, the Ewelme and the Little Avon fulfilled the requirements, with many mills around Stroud and Dursley. In Wiltshire, there was another group of mills along the Calne Water. Some of the early fulling mills were sited at local node points, where different users could reach them, especially in upland districts. This was one of the reasons why Sowerby Bridge became one of the first places in Yorkshire to have a fulling mill. The fulling mill was introduced to Wales in 1295, with seventy-one being built over the next forty years. It made such an impact locally that the Welsh word for

Ground-floor operation of the fulling mill at Helmshore, Lancashire.

This late medieval mill at West Harnham, near Salisbury, was originally used for fulling. This photograph dates from the 1950s. It is now part of a hotel. (Museum of English Rural Life)

the mill, *pandy*, became incorporated in many place names.

The major exception to the expansion was East Anglia, which was the third major clothing district in England, yet contained far fewer fulling mills than other areas.[92]

Many of the new fulling mills of the late fourteenth century were converted from corn mills that were suffering from lack of demand after the Black Death. Conversion was a relatively simple operation – the fulling stocks occupied the ground floor, worked by a camshaft from the adjacent waterwheel, whereas the corn-grinding stones were on the first floor. Other fulling mills shared premises with a corn mill, in which case a single-storey extension to the building might be added to accommodate the fulling stocks.

The subsequent fortunes of water-powered fulling fluctuated along with those of the different branches of the woollen industry, and the relative value placed upon this process. It did not necessarily replace the hand processes: indeed, there are photographs of people working cloth by hand in some parts of Scotland and Ireland in the twentieth century. In the Middle Ages hand processes continued in use for the manufacture of East Anglian broadcloth while West Country clothiers took up water power. In East Anglia there was greater competition from cornmilling, and because fulling was just a small part of the manufacturing process for cloth, clothiers were unable or unwilling to outbid the corn millers for the mills. The relative importance of fulling in the manufacture of cloth is one explanation for the ease and frequency with which mills changed use. If it paid better to grind corn, make paper or crush stones than to full cloth, out came the fulling stocks and in went the other equipment; the reverse happened when fulling was a good economic proposition. Similar considerations resulted in those instances where fulling and grist mills shared a site. These fluctuations make it very difficult to have a clear idea of total numbers of fulling mills, which can be estimated only by looking at local generalizations.

Where fulling mills did become established, as in the south-west, they were often expanded with the addition of other processes. One of the first was the gig mill, which raised the nap of finished cloth; the water-powered mill was adopted from the sixteenth century onwards. The strength of water-powered fulling in the West Country and the early growth in the number of mills were associated with the manufacture of undyed broadcloth. That lasted

Loudwater Mill, Buckinghamshire, is on one of those sites that has been through a number of uses, from fulling to paper and finally corn mill, the form in which it was photographed at the beginning of the twentieth century. (Museum of English Rural Life)

until the policies of James I early in the seventeenth century led to a serious decline in the trade. The new cloths introduced to replace the old broadcloth required less fulling, and the number of mills in this region declined. It was a similar story in East Anglia, where broadcloth manufacture declined after the outbreak of the Thirty Years War in Europe in 1618. The new types of cloth – the says, bayes and shalloons, known collectively as 'new draperies' – which increased in importance in the region, required little or no fulling. All of these developments led to a decline in the number of fulling mills during the first half of the seventeenth century. Many were converted to other uses. Colthrop Mill on the River Kennet in Berkshire was fairly typical. Having started out as a corn mill in the Middle Ages, it became a fulling mill in the early sixteenth century as the cloth industry of Newbury grew. A reversal of fortunes for the local industry saw the mill return to corn-grinding in the early seventeenth century, and continued in this activity until being converted to a paper mill in the 1740s. Further down the Kennet, at Reading, the fulling mills had gone by the end of the seventeenth century.[93]

In northern England, some of the new gentry purchasers of monastic lands after the dissolution were keen to promote the development of their estates, and the provision of fulling mills was one means they adopted. The Ramsdens, later prominent in the development of Huddersfield, were one family engaged in this work.[94] There was renewed interest in fulling mills in the West Riding and Pennine Lancashire from the late seventeenth century onwards, when growth revived in some branches of the woollen industry. Old fulling mills were brought back into operation and a number of new mills were built. By the 1770s, more than a hundred fulling mills were at work in the West Riding, and this number had risen to 197 by 1800.[95] Baines's directory of Lancashire of 1825 referred to twelve fulling mills in Rossendale, ten in Rochdale and seven around Bury. Many fulling mills continued in use throughout the nineteenth century and into the twentieth. As well as fulling woollen cloth, the stocks were used to beat skins in preparation for the manufacture of chamois leather. A tannery at Torrington, Devon, was doing this in the 1880s.[96]

The process of re-use of mill sites means that fulling mills built before the eighteenth century are not readily identified. Those that continued in textile production were almost all incorporated into new and larger mills in the eighteenth or nineteenth centuries. Sometimes, the only reminder of a fulling mill is in a local street name.

The preserved Higher Mill at Helmshore in Lancashire is of the later generation of fulling mills. It was built in 1795 and the fulling stocks of the early nineteenth century and a waterwheel of about 1830 survive. Some buildings of those fulling mills converted to grind corn survive occasionally, as at Dursley and Painswick.[97]

Apart from the monastic estates, the builders of the medieval fulling mills were usually the lay landowners. Where manorial rights were strong these lords exercised the same controls over fulling mills as they did over the corn mills. The same types of dispute over the rights might also arise. However, the rise of powered fulling after the Black Death came at a time when estates were loosening manorial controls. Most mills were leased, rather than directly managed, and the lessee was usually a clothier or group of clothiers, merchants who took responsibility for the cloth-finishing processes. In some places, independent craftsmen-weavers ran the

Washford Mill, one of Congleton's silk mills of the mid-nineteenth century. It drew its power from the River Dane, continuing to use it well into the twentieth century.

fulling mills, as was the case in the district around Wilton and Salisbury.⁹⁸

SILK AND THE FIRST FACTORY

In the early eighteenth century a water-powered silk mill was built at Derby. This assumed a significance out of proportion to the size of its industry, for in the scale of its operation it was the precursor of the factories of the Industrial Revolution.

Silk cloth had been produced since the late Middle Ages, using raw silk imported from Italy and the Near East. The English silk industry was on the small scale, both for the production of yarn ('throwing' was the term used rather than spinning) and the weaving of the cloth. There were some notable concentrations of silk weavers, in Coventry, Spitalfields and parts of East Anglia, for example. Simple throwing machines developed in the Middle Ages could be used in the production of yarn. Although they could be operated under power, most were small enough to be worked by hand, and that was certainly how it was done in England. In Italy, Europe's leading producer of silk, water power was applied to silk throwing much earlier, possibly as early as the thirteenth century. It was certainly well established by the seventeenth century, with improved machines, and this was the technology introduced in Derby.

Thomas Cotchett was the man who built the first water-powered mill at Derby for throwing silk of the highest quality, known as 'organzine'. The mill was on an island site on the River Derwent created by the millrace for the town corn mill, which now also fed Cotchett's mill alongside. Accounts vary, but it seems that George Sorocold installed the millwork some time around 1702–04. Cotchett's venture did not have a long life, but among those he employed, perhaps as an apprentice, perhaps as an assistant, was one John Lombe, who came from a family of worsted weavers. In 1716 Lombe engaged in some industrial espionage to discover the most up-to-date techniques of the Italian silk industry. He gained a British patent in 1718 for the throwing machines he designed based on his observations and, together with his half-brother Thomas, built a mill next to Cotchett's. After John Lombe died, in 1722, Thomas took sole charge of the mill. It was big: five storeys high, 110 feet long and 39 feet wide, and contained twelve throwing and seventy-eight winding machines, employing 300 people. George Sorocold also built the wheel for Lombe's mill – it was an undershot wheel of 23 feet in diameter, on the outside wall of the building, not enclosed in a pit or wheel-house. Power was transmitted from the waterwheel by a horizontal line shaft, which ran underneath the building. At intervals along its length, toothed gears took the drive to vertical shafts running up to the machinery floors above.

Lombe's mill soon gained celebrity, praised by Daniel Defoe as 'the only one of its kind in England'. Defoe also recounted what was probably a well-known tale at the time. It seems that George Sorocold, showing some visitors his mill wheel, slipped into the stream, was swept round the wheel and deposited in the tailrace, apparently none the worse for the experience.⁹⁹

Lombe's mill succeeded in undercutting the price of Italian organzine. Although it did not stay in business for long after the expiry of Lombe's patent, in 1732, it did set the example that was followed by more water-powered silk-throwing mills to produce organzine. The first was at Logwood Mill, Stockport, in the 1730s, followed by the Button Mill at Macclesfield in 1744, and the Old Mill, Congleton,

The silk mill at Whitchurch, Hampshire.

in 1753, which, at five storeys and thirty bays, was on a similar scale to Lombe's. The silk industry expanded during the late eighteenth and early nineteenth centuries, with many mills producing organzine and the lower-quality yarn called 'tram'. By the end of the century imports of raw silk were double what they had been at its beginning. Macclesfield and Congleton were the principal centres: by 1765, east Cheshire accounted for a third of the English silk-throwing industry. Water power was important, more so in Congleton; in Macclesfield, the streams were more limited in power and rights to use them more restricted.[100]

The industry was also being developed in southern counties, partly in an attempt to offset the decline of the older woollen manufacture in such places as north Hampshire. By the end of the eighteenth century there were silk mills at Overton, Whitchurch and Winchester. Often they had been converted from other uses. Twyford Mill in Berkshire, on a medieval milling site, was taken over by two brothers from Macclesfield, Thomas and George Billings, in 1810, and converted to silk processing. This mill continued in use under a succession of owners until it was destroyed by fire in 1891. The Victoria Mill on the River Test at Whitchurch is believed to have been a fulling mill. By the end of the nineteenth century it was almost the sole survivor of southern silk manufacture. It continued to operate commercially until the 1980s, and survives now as a working museum.[101]

The successful introduction of water power to the silk industry made it the branch of the textile industry that was second in importance to wool, until the mechanization of cotton-spinning that came with the Industrial Revolution. The linen industry lagged behind at this stage, but in the early eighteenth century the introduction of water power for scutching marked a step on the path of its expansion. Scutching was a preparatory process that broke down the fibrous material of the flax in preparation for being woven into linen cloth. Water power was also applied to finishing processes after the weaving: washing and beetling (the beating of cloth with timber stamps to produce a fine sheen).

Water-powered mills for preparing dyestuffs were being operated by the mid-sixteenth century. There was one in 1580 at Wandsworth, then outside London, on the River Wandle. Later expansion resulted in a windmill being added to the building in the eighteenth century.[102]

METAL INDUSTRIES

If the existence of water power in the production of metals in Roman times is conjectural, there is far more concrete evidence from the late Middle Ages. Most comes from the fourteenth and fifteenth centuries, when it is clear that water power was being introduced extensively. Demand for military use was one stimulus to greater production from the late fourteenth century onwards. It especially boosted the iron industry of the Weald, which was able to produce hard iron for cannon. Iron manufacture was the largest of the metal industries in Britain and, in consequence, the greatest user of water power, but all metalworking industries took up water power from the late medieval period onwards.

The method of making iron inherited from the ancient world involved the direct smelting of ore at temperatures that could reach 1200°C. Two processes lent themselves to mechanization: first, the laborious task of operating bellows for the furnace. In medieval England these furnaces were known as 'bloomeries', and the product was a 'bloom' of wrought iron. The second process that was mechanized was the repeated hammering and re-heating of the bloom to remove impurities from the wrought iron. The use of powered bellows and trip hammers represented a great saving of manual effort: records of two sixteenth-century bloomeries, one near Sheffield using water power, the other at Llantrisant with no power, show the latter using three times the labour on the bellows.[103]

Water power was introduced to these processes during the central Middle Ages, following earlier experience in France, Germany and Bohemia. Speculation that the record in Domesday of the payment of rent in iron blooms for four mills in Somerset represented water-powered ironworks remains just that. More certain evidence is provided by the water-powered forge that has been excavated at Bordesley Abbey, Worcestershire. Some of its remains have been dated to the late twelfth century, and it was possibly, therefore, the earliest forge to use water power. It continued in use, being rebuilt at least twice, until the fourteenth century. The wheel was undershot and drove bellows and a trip hammer. There is another record of a leat being built c.1200 at Kirkstall Abbey, near Leeds, apparently to feed a forge.[104] Several other sites of water-powered furnaces or hammer mills have been identified, including one at Rievaulx Abbey, and another at Chingley in the Weald of Kent. By the thirteenth century, the iron industry of the Weald, the Forest of Dean and northern England was using water power to drive the bellows for the furnaces. Waterwheels to drive metalworking hammers are known from the mid-fourteenth century. Powered stamping presses to break up the ore appeared at about the same time. By the seventeenth century water-powered hammers were common in iron-making. In Cumbria, for example, there were several forges with water-powered trip hammers.[105]

The 'indirect' method of making iron, by which the ore was smelted in a blast furnace to produce pig iron, which had subsequently to be re-worked as cast iron, was developed in continental Europe in the early fifteenth century. It was introduced to Britain in 1496, at Newbridge in Ashdown Forest. It needed greater power than the bloomery furnaces for the two principal operations of working the bellows and the hammers, and water was by now the obvious source for this. Bellows were operated by cams off the wheelshaft that engaged with the arm of the bellows. The first few blast furnaces were built in the Weald, then the leading iron-producing centre, and well placed to invest in new technology. When they were introduced, however, demand for iron was sluggish. Bloomery furnaces remained common in much of the country until well in the seventeenth century, while blast furnaces grew in numbers very slowly until the 1540s: there were none in the Forest of Dean, for example, before this time. During the second half of the sixteenth century they became more widespread, and began to be found in such places as Shropshire, Staffordshire and Glamorgan. Even so, there were still only three in the Forest of Dean by 1620. The Wealden industry was still the dominant user of this technology, with forty-nine out of the seventy-three blast furnaces in England in 1600. The charcoal-iron industry was introduced to Ulster during the seventeenth century, and water

A water-powered trip hammer at Holbeam Mill, Devon, an edge-tool mill that last worked in the 1940s. (Mills Archive Trust)

power largely determined the siting of these works.[106]

Water power was further deployed in iron manufacture, being used for bellows at the fineries where pig iron was re-heated for moulding, and for operating forging hammers. Such fineries and forges were common in both the Weald and the Forest of Dean by the early seventeenth century. The iron could be rolled into sheets or cut (slit) into bars, and water power was applied to both processes. Introduced into France in the fifteenth century, the first slitting mill known in Britain was not working until 1590, at Dartford, Kent. Rolling mills were water-powered in Britain from the seventeenth century.

Other metallurgical industries used water power increasingly for washing and crushing ores, and to operate tilt hammers to forge the finished products. Water power was also introduced to the tin industry of south-west England, although precisely when is uncertain. In one lease of the mid-twelfth century from Devon, the lessor excludes from the grant of land a mill, leat and mines of tin, but whether mill and mines were connected is not clear. It is quite likely that water power was used for tin mills during the thirteenth century. The ground is surer with the introduction of the blowing house, or smelting mill, in the mid-fourteenth century, for which the bellows were operated by waterwheel. This allowed temperatures in the furnace to be much higher.[107]

The tin industry of Cornwall and Devon had one of its boom periods in the late medieval period, starting in the early fourteenth century, and continuing, after a setback following the Black Death, through to the sixteenth century. Demand was sufficiently high to justify investment in water power for the blowing mills and crushing mills, and for cleaning the ore in the smelting works in both Cornwall and Devon. There are many references to 'blowing houses' in contemporary sources. Stamping presses (also known as 'knacking', 'knocking' or 'clash mills'), which crushed the ore, were also built in quantity. More than 150 were built in Cornwall and about sixty on Dartmoor. This activity gave the south-west the greatest concentration of industrial water power in late-medieval Britain. Ten per cent of the mills in this region were for uses other than grinding grain in 1400, rising to 25 per cent by 1480 and 42.7 per cent by 1540. By the end of the sixteenth century this tin boom was passing as brass replaced pewter in public affections. Output continued at much lower quantities throughout the seventeenth century, until demand built up again during the eighteenth. The rapid pace of this decline means that little evidence survives on the ground of this phase in the industry's activity.[108]

The Society of Mineral and Battery Works, one of the joint-stock companies set up under royal licence in the late sixteenth century, established water-powered brass and wire-drawing works at Tintern. At the same time the Bristol area was becoming noted for its copper and brassware. Water-powered hammers were used in many of the works. One, the Baptist Mills on the River Frome, established in 1702, included Abraham Darby among its partners, until he left to move to Coalbrookdale. The Bristol Brass Company took a lease of a fulling mill at Saltford on the Avon in 1721, where it continued to use water power for rolling mills and for working the utensils (battery work) until 1908. There were brass-ware mills on the Thames in the Marlow area, Daniel Defoe reported, but they were bankrupted by the South Sea Bubble in the 1720s.[109]

Water power was applied to the manufacture of metal tools, especially edge tools, in the late Middle Ages. There were two fairly large blade mills at Winchester, but at this time they were barely profitable, and the rent paid to the lord of the manor was on the low side.[110] During the sixteenth century, water-powered blade mills for grinding edge tools and cutlery became more numerous, and it was from this time that the areas most closely associated with these trades became established, each with its range of specialities. The metal trades of the West Midlands trace their origins to this time: Bromford Mill in Staffordshire was among the first blade mills to be established in the sixteenth century. More rapid expansion followed in the seventeenth century, so that by 1700 the River Tame and its tributaries had a particular concentration of blade mills. Individual mills were small, and required little power. Further growth in these trades, together with the addition of rolling and slitting mills, which were built to satisfy demand for sheet metal, resulted in almost all the water-power sites within 5 miles of Birmingham being in use by 1750.[111]

In Worcestershire another specialized local business using water power grew up, with about eighteen mills being employed in the production of needles. Most were on the River Arrow, and a few were further afield, on the River Alne and at Bromsgrove. Formerly used as forge mills and corn mills, these

The needle mill at Redditch, Worcestershire.

mills were employed in the pointing and scouring (polishing) of needles. The first was Forge Mill, Redditch, converted to a needle mill about 1729. It was also the last at work, closing in 1958, and now preserved. This was a particular localized use of water power, prompted perhaps by the availability of mills for conversion and the high quality of needles produced with water power. Needle makers elsewhere, at Long Crendon, Buckinghamshire, for example, kept to hand processes.[112]

It was a similar story in the cutlery trades. The first water-powered grinding mill for cutlery appeared on the River Sheaf, Sheffield, in 1496, and a few more followed shortly afterwards. Rapid expansion of the tool and cutlery industries from the seventeenth century onwards filled the rivers around Sheffield with forges, edge mills, wire works, and slitting and rolling mills, alongside corn mills and paper mills. In 1794 a local directory listed 111 water-powered cutlers' wheels.[113]

The south-west and north-west of England – at Dalston, Cleator and in Furness, for example – were two more areas where water-powered workshops were established during the seventeenth and eighteenth centuries. These were making spades, scythe blades and plough shares.[114] Other mills making domestic products elsewhere included a thimble-making mill at Marlow, Buckinghamshire, in the early eighteenth century.

MINING

Mining requires pumps to drain excess water away, as well as hoists for miners, their equipment and the produce of the mine. As mines became deeper and more complex, they outgrew the power that horse gears could supply, and turned increasingly to water power. This developed most rapidly in Germany and central Europe, so that by the mid-sixteenth century the use of water power was common, in increasingly elaborate arrangements. Technical literature was produced to describe the workings of the mines and their power, most notably *De Re Metallica*, published by Georgius Agricola in 1566.

Although water power was already being used for processing the ore, it was not employed for mine pumps in Britain until the foundation of the Company of Mines Royal in the 1560s. This group of German miners, who were granted a royal licence to mine copper in Cumberland, brought with them many techniques from the continent to their mines at Newlands, near Keswick, including the use of water power for pumping and for stamping and crushing ore. Their example was followed by mines elsewhere: by the late seventeenth century the lead and silver mines in Derbyshire, Yorkshire and Scotland were all turning to water power for pumping, driving stamping presses and bellows.

Water power reached the British coal industry in the late sixteenth century. At Wollaton in Nottinghamshire there was a 'coal mill' in 1580, comprising a wheel that drove pumps using buckets or chain engines to raise water from the mine. By the middle of the next century this type of pump was common, although there was a limit of 90–100 feet to the depth from which it could draw water. As deeper mines were opened, more elaborate arrangements of wheels and pumps were constructed, from the mid-seventeenth century onwards. At Ravensworth, County Durham, three wheels of 24 feet in diameter worked in series, and by the end of the seventeenth century Lumley colliery had as many as nine wheels. The use of reciprocating pumps instead of the chain engine enabled the water pumps to reach a depth of 240 feet.

Waterwheels were turned to the task of draining the tin mines of Cornwall. As demand revived in the eighteenth century, entrepreneurs in Cornwall turned to the new power of steam for drainage pumps. Savery's engine of 1698 was followed by a much better one developed by Newcomen, introduced in 1712. Ultimately, steam engines proved to be more effective at deep pumping, but the water pumps were certainly not replaced immediately. Being relatively cheap to run, many were kept going until the costs of maintenance became prohibitive, often well into the nineteenth century.

PAPER-MAKING

Paper-making requires power to pulp the rags, wood or other raw material. When paper was introduced to Europe early in medieval times, the pulping was the work of hand or foot power. Water power was first applied to the process in Spain in the mid-twelfth century, and paper-manufacturing works have ever since been known as 'mills'. It was not until 1495 that the first paper mill was recorded in England; this was John Tate's Sele Mill near Hertford. By that time water power was established in paper-making. Although Tate supplied paper to the leading English printers, William Caxton and Wynkin de Worde, competition from foreign makers was evidently strong, for his mill did not survive long, if at all, after his death in 1507. The paper-making industry began to be established in Britain in the second half of the sixteenth century, but even then there were many short-lived conversions of corn mills to paper mills. Sopwell Mill, St Albans, was one example, making paper in 1649 but reverting to corn by 1691.[115]

Part of the problem was that mills were competing for sites with other expanding industries, while competition from overseas suppliers of paper was strong. There were opportunities in some places: in Kent the decline of a local textile industry left a number of fulling mills available for conversion. One of the most important mills in the paper industry, Turkey Mill, near Maidstone, was established in this way in 1670. The industry was growing strongly at this time, and continued to do so into the early

Early Industrial Uses of Water Power

Bere Mill, Laverstoke, Hampshire, where Henri Portal first set up as a maker of paper. (Museum of English Rural Life)

eighteenth century: the number of paper mills in England reached 106 by 1690 and nearly 200 by 1710. Metropolitan demand can be seen in the large number of mills along southern rivers, in Hertfordshire and Kent, along the Len, for example, and the Wye in Buckinghamshire. West Mills at Newbury became a paper mill at this time, expanding to become one of the bigger eighteenth-century producers. Competition for mill sites could still be strong, however, and paper mills on the River Derwent in Derbyshire at Masson and Darley Dale were priced out by the new cotton industry. The demands of an industrializing nation provided the foundation for continued growth. Water provided the power universally until the beginning of the nineteenth century, when steam started to become established. At the same time, consolidation of the industry was beginning.[116]

Pulping the rags used the same basic action as fulling, and it was not unusual for a paper mill to be converted from a fulling mill, or built on the site of one. Some, such as the paper mill opened at Milnthorpe in Cumberland about 1700, were formerly ironworks, with water-powered trip hammers.[117] There were some that shared premises with other mills, for most water-powered paper mills of the seventeenth and early eighteenth centuries were small. The most usual neighbour was a corn mill, as at Wansford, Northamptonshire and Cattershall on the River Wey. At Sutton Courtney, Berkshire, a paper mill and four corn mills were located on the same site at the end of the seventeenth century. At Exwick, near Exeter, the same premises were shared by a fulling mill, a corn mill and a paper mill.[118]

Paper-making was scattered throughout the country. The principal requirement was a supply of good clear water, especially for the production of higher-quality white papers. Satisfying these conditions meant that the mills tended to congregate at certain localities, at Croxley, near Watford, for example. The clear chalk streams of Hampshire were an ideal location for mills engaged in the manufacture of speciality papers. Paper for banknotes was a particular specialism of this district and became the basis of one family's fortune. Henri Portal, of a Huguenot family, took a lease on Bere Mill, Whitchurch, in 1712. His business, specializing in very fine paper, expanded enough for him to take on the additional lease of Laverstoke Mill a little further upstream towards Overton. He rebuilt this old manorial corn mill for paper-making, and it became the centre of the business. The contract for producing paper for Bank of England notes was awarded in 1724 and held into the twentieth century. Alton was another centre for paper-making in Hampshire, retaining mills throughout the nineteenth century. The principal mill here was operated by the Spicer family in the late nineteenth century, mainly making fine-quality writing papers, many of which involved hand processes.[119]

At the end of the seventeenth century a new pulping machine was introduced, the hollander, or beater engine. This worked with a chopping and slicing action instead of the pounding of the hammers that had been used before. It was more efficient, making much better use of the water power – one waterwheel could more easily drive several hollanders than a row of stamping presses – and the new machine was taken up widely during the early eighteenth century. At about the same time, water-powered glazing presses were introduced, which applied a smooth finish to the paper. The new machinery increased the capital needed for paper-making, and often the scale of production. In the 1740s Joseph Sexton spent a considerable sum, more than £1,000, equipping his mill in Limerick, while some manufacturers, including John Spilman of Dartford, one of the most celebrated paper makers, were becoming major employers.[120]

SEED-CRUSHING

Crushing oil-bearing seeds using edge running stones became a powered process in the sixteenth century. The oil produced was an important commodity in the finishing processes for woollen cloth, and was used also for lighting and lubrication. Most of the seeds were imported from the Low Countries through the ports of eastern England. An extensive seed-crushing industry built up at Hull, which became the principal port, and at Gainsborough. The oil mills at both towns were windmills. Water power was used elsewhere, however, sometimes associated with local cultivation of rape and hemp, the principal sources of the oils. Benedict Webb, a sixteenth-century clothier, built a mill at Kingswood, north Somerset, where he ground rapeseed from his own land. There were other mills at Rotherhithe on the Thames and at Caerleon in south Wales, while Daniel Defoe mentioned another Thames mill at Marlow that was crushing flax and rape seed. In the Tame valley, Staffordshire, a number of watermills for seed-crushing were at work in the seventeenth century, and continued into the early eighteenth century. There were some in Yorkshire, at Great Ayton and Fryup, for example, which operated in the eighteenth to mid-nineteenth centuries. Water-powered seed mills were introduced to Scotland in the early eighteenth century.[121]

GUNPOWDER

When the manufacture of gunpowder began in Britain, in the fourteenth century, it is probable that most of the processes of grinding the ingredients – saltpetre, charcoal and sulphur – into the powder were carried out by hand or animal power. Grinding on a larger scale used edge-runner stones, for which a simple drive could be taken from a vertical waterwheel. The first positive references to water-

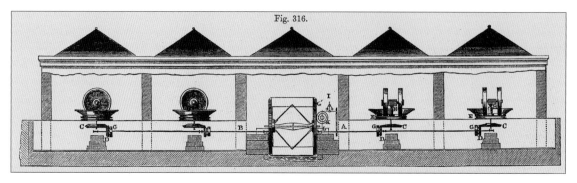

Water-powered edge-runner stones for mixing gunpowder. A diagram from Sir William Fairbairn's treatise on mills of a wheel installed at Waltham Abbey. (Museum of English Rural Life)

powered mills date to the mid-sixteenth century. These mills were producing gunpowder for ordnance, and most were clustered in the London area. The Lea Valley was one centre, another was the Faversham area of Kent. In Surrey there were mills at Wotton and Abinger, which by the 1640s were superseded by larger works further down the Tillingbourne valley at Chilworth, in which the East India Company had an interest. Many of these mills were on sites converted from other uses. By the late seventeenth century all gunpowder mills were grinding by water power.[122]

The stage in the manufacture at which all the ingredients were mixed together under pressure, and water added, was known as 'incorporating'. When this was done by hand, a mortar and pestle were used. Stamping presses operated by horse engine allowed larger quantities to be mixed. The risk of explosion was high and this was a deterrent to the introduction of water-powered presses, which would have increased production further. In the 1680s edge-runner stones were applied to the incorporating process at the mills at Woburn in Bedfordshire. These were much safer than the stamping presses and enabled water power to be used. For safety reasons, legislation in the eighteenth century abolished the use of stamping presses.

Waltham Abbey gunpowder mills were founded in the 1660s on the site of a water fulling mill on the River Lea in Essex. They became the most important and extensive of the ordnance works, and long-lived, for they stayed open right up to the 1980s, having turned from making gunpowder to the new chemical explosives at the end of the nineteenth century. Water power was extensively used until the 1850s, when the first steam-powered mills were built to meet increased demand during the Crimean War. Water power was so important to these mills that their owner, Richard Walton, opposed the New River Bill introduced in 1739 to improve navigation of the Lea, because he was anxious about how his water supply might be affected. In the 1770s additional water-powered mills were still being added to the site to replace horse mills.

A market for gunpowder for civil purposes developed during the eighteenth century, mainly for mining and quarrying. The slate quarries in the Lake District provided the demand that prompted John Wakefield to start the first gunpowder mills in the north of England, at Sedgwick, near Kendal, in 1764. The industry grew, supplying the mines and quarries of northern England, as well as overseas markets. Mills along the rivers Kent, Leven and Brathay, and some of their tributaries, constituted an industry that reached its peak between 1860 and 1918, drawing in many workers displaced from declining bobbin manufacture. Although steam power was taken up during the second half of the nineteenth century, water power was still important, the vertical wheels replaced sometimes by turbines. Steam might be used for the incorporating, but with workshops necessarily spaced far apart over the site of the works, it was neither convenient nor safe to have a number of small steam engines and boilers. For other processes, therefore, water power was often preferred. All the mills of Westmorland came into the ownership of ICI when it was formed in 1926, and it was decided to concentrate production of explosives for civil use at Ardbeer, Ayrshire. The last of the north-western mills closed in 1937.[123]

WATERWORKS

Using the power of water to raise and pump water has been one of the most enduring applications of the technology. It was also one of the earliest, with the noria wheel, which was designed for irrigation systems. Water-driven pumps were used on farms and estates until the establishment of mains-water supplies largely rendered them unnecessary.

In the early modern period, waterwheels were employed for the public water supply and for drainage. The growth of towns led to a greater demand for water and for a more regular supply of it. An early example of artificial water supply was the installation of a waterwheel below London Bridge in 1582, which drove a pump to lift water from the Thames into a reservoir, from which it could be distributed in the city. The London Bridge Water Works lasted until 1822, and achieved some renown. Similar works were built at Marly, near Paris, in the

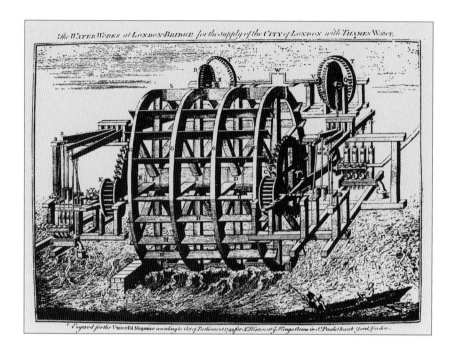

A depiction of Sorocold's waterwheel for the London Bridge Water Works from The Universal Magazine *of 1749. (Museum of English Rural Life)*

1680s, using large waterwheels to pump water from the river to supply the gardens and palaces at Versailles and Trianon. There was some rivalry in contemporary comment as to whether London or Paris had the greater installation.

The works at London Bridge were rebuilt and enlarged several times, until eventually there were five waterwheels working the pumps. The first major rebuilding followed the destruction of the works by the Great Fire of London in 1666. Major reconstruction work was undertaken, started in 1701 by George Sorocold, who was establishing himself as one of the foremost hydraulic engineers in Britain. He had already built water-powered pumping systems to supply water for a number of towns and cities. Macclesfield was the first, where he employed a pumping system driven by waterwheel, and he did similar work in Derby (his home town), Bristol, Norwich and Newcastle upon Tyne. His pumping system was occasionally put to work in mines, one being at Alloa, where he installed the equipment in 1710. He had also undertaken work for Thomas Cotchett's silk factory at Derby, and was subsequently the engineer for the larger factory built by the Lombe brothers.[124]

Sorocold's work at London Bridge involved the replacement of much of the wood in the wheel and gearing by cast iron, these waterworks being one of the first major users of iron for millwork. The four wheels built at this time were 14 feet wide. In the 1760s further substantial work was carried out following pressure from the users of the river, who wanted more open navigation through the bridge – with its waterwheels, it acted as a dam. John Smeaton undertook work for the water company at this time, designing a new wheel for the fifth arch of the bridge, larger in diameter, at 32 feet, than the previous biggest of 24 feet 6 inches. Smeaton's wheel had improved float boards, and the gearing was also improved, resulting in a substantial increase in power output. Twenty years later, between 1786 and 1796, John Torr Foulds, the senior millwright and engineer of the water company, undertook another complete reconstruction of the wheels and associated works. The pressure for improvements to the navigation of the Thames was, however, building up again, and, although a new iron wheel was commissioned in 1819, the waterworks were abolished by Act of Parliament in 1822.[125]

7 Water Power and the Industrial Revolution

With the various technical developments the use of water power was extended in all directions, far beyond the flour-milling that predominated at the time of Domesday Book. Decline after the Black Death had been more than made good, and the number of waterwheels had grown enormously since the end of the Middle Ages. John Evelyn, owner of Wotton estate in Surrey, remarked of his local river, the Tillingbourne, 'I do not remember to have seen such a variety of mills and works upon so narrow a brook and in so little a compass, there being mills for corn, cloth, brass, iron, powder, etc.' Evelyn was himself involved in the use of water power, having interests in the local gunpowder works.[126]

Quite how many waterwheels there were is impossible to tell, but it has been suggested that there could have been as many as 20,000 watermills in Britain by the late eighteenth century. Local studies give an indication of how extensive water power had become. In the county of Westmorland the number of mills grew from twenty-five in the fifteenth century to more than ninety by 1850, a rate of growth which gave the county a greater amount of water power per head of population than Birmingham. The use of water power around Sheffield was in its infancy in 1500, but by the late eighteenth century more than a hundred mills were active, most of them in the metal trades. There were sixty watermills on 3 miles of the River Mersey below Manchester in 1700. At the same time on the small River Ecclesbourne in Derbyshire, nine mills crowded on to a river 8 miles long, draining lead mines, crushing ores, preparing dyes and crushing bones.[127]

All this was happening before the next (and arguably the greatest) phase in the employment of water power – to drive the machinery for the new factories of the Industrial Revolution. Steam power is normally associated with industrialization, but in its early phase, until the late eighteenth century, the waterwheel was the only practical way of driving machinery in all but the smallest textile mills or ironworks, where horse engines might be used. The first steam engines had been built by Thomas Savery at the end of the seventeenth century, followed by Thomas Newcomen with his engine in 1712. These engines were limited in their application because they could operate only in the vertical plane; as a consequence, they were used mainly for pumping water from mines. They used steam at low pressure, usually little above natural air pressure, and were thus commonly referred to as 'atmospheric engines'. Their low power meant that the waterwheel was more than a match. James Watt made great improvements to steam engines, but he continued to use low-pressure steam; his engines were still of modest power, and it was not until the 1780s that he succeeded in taking rotary drive from the vertical cylinder.

For the industrialists of the eighteenth century there was little alternative to water power to drive their new cotton mills. Their demand for power spurred on development in the technology of waterwheels to such an extent that water power continued to be a practical proposition for some time after steam engines had become more competitive in efficiency and cost. The early stages of the Industrial Revolution in Britain thus were mainly powered by water.

Large-scale production in factories was at the heart of the new industrialism of the eighteenth century. Its origins can be traced back to the water-powered silk mills at Derby, and the mills in Cheshire in the mid-eighteenth century. The main focus, however, was cotton-spinning, followed a little later by developments in the woollen and worsted industries. The growth of these industries led to an expansion in the use of water power and a development of the technology to increase its efficiency. Sites were quite readily available, at least initially. In the north of England, where most of the development of these industries took place, there were unused sites on many rivers, as well as sites that had been used by other industries, or by mining that had declined or moved away, which were available for re-use. A number of eighteenth-century cotton mills were built at sites that had been used by the iron industry, including Strutt's mill at Milford in Derbyshire, and Birch & Co at Backbarrow, in Furness.[128] A number of corn mills were converted to textile mills, especially in the woollen industry of the West Riding and in Wales. Old corn mills were available for conversion partly because greater efficiency could concentrate production on fewer mills, and partly because farming in the north and in Wales was starting to focus more on livestock than cereals. In southern England, the reverse process was taking place: as the textile industry migrated north, many mill sites were freed for conversion to corn-grinding.[129]

The ready availability of water-power sites was a short-lived phenomenon. The expansion of the textile industries, and of a number of other activities all wanting water power, soon began to put pressure on the rivers. Cotton masters had to look further afield for a site for their new mills – to the Lancashire coast or north Wales, for example. Yorkshire woollen mills moved further up the Pennine valleys. This was already happening by the 1780s, and ultimately pressure on the available water supply became a major consideration in the choice between water and steam power.

THE COTTON INDUSTRY

In 1767–8, Richard Arkwright invented a mechanical frame for spinning cotton. He was awarded a patent for it in 1769 and the following year he joined with other partners to set up a factory at Nottingham, where his spinning frame was to be worked by horse engine. However, even before the factory was finished (it did not open until 1772), Arkwright was looking to expand beyond the limitations of horse power. As he turned to waterwheels, his spinning machine quickly became known as the 'water frame'. It was a name that encapsulated the nature of the eighteenth-century Industrial Revolution – it was powered mainly by water.

Richard Arkwright was not the first to use water power for textile manufacture. The silk industry was an important precursor. The Lombes had had their silk works at Derby for fifty years and there were other silk works using water by the 1760s at Congleton, Macclesfield, Stockport and Sheffield. In cotton-spinning, the first steps in the use of water power had been taken by Lewis Paul in Northampton and John Wyatt in the 1740s and 1750s. However, Arkwright's spinning frame was one of a number of new machines that placed the cotton industry at the heart of industrialization in the eighteenth century. The machines for carding and spinning could be worked by hand, and many were. More were operated by horse gears, which were often adequate in small textile works at the beginning of the nineteenth century. Horse engines had limitations, however, and this was what prompted Richard Arkwright to build a bigger, water-powered mill. He sought additional capital, taking Jedediah Strutt and Samuel Need, wealthy hosiers, into partnership, and in 1771 moved from Nottingham, where the slow-moving rivers were lacking in power, to Cromford in Derbyshire. There the partners entered into a lease that granted them rights to use water from the Bonsall Brook and Cromford Sough, a drainage channel from the local lead mines, to power mills. On this site Arkwright built a new mill using water to power his mechanical spinning frames.[130]

Cromford was not, perhaps, the most obvious choice of site for a new textile factory. It was relatively isolated from the major markets, being 14 miles from Derby, 26 from Nottingham, and 45 from Manchester, and was not on the turnpike road to any of them, although packhorse routes did pass through. The road that is now the main A6 through the Derwent Valley to Matlock was not built until 1820. The village was small, but, this being a mining area, mainly for lead, there was the prospect of a reasonable supply of local labour. This was one of the things that attracted Arkwright, although he did have to advertise in newspapers in Derby for additional workers. He seems to have regarded the local workers as more reliable than those of his native Lancashire – fear of machine wreckers was said to be one of Arkwright's main concerns. He was also keen to protect his patent and his industrial methods, and the security afforded by this relatively remote site may also have been a consideration. Over all of that was the water: according to Arkwright, the little Bonsall Brook and Cromford Sough offered 'a remarkable fine stream of water', and the Sough was reported never to freeze. Reliability was what Arkwright wanted. He was, it is said, initially suspicious of the Derwent, which appeared to be capricious.

Just as importantly, probably, he was also able to obtain clear rights to the streams. Although the Derwent offered plentiful potential for water power, without the complications of navigation interests, there was competition for water rights from mining groups, paper-makers and others. At Masson, where Arkwright later built a mill, the rights to the water were in the hands of others, Robert Shore and George White, who acquired them in 1771. It is also likely that Arkwright, this early in his career and short of capital, could not afford the more extensive engineering work that would be needed to tap into the Derwent's power.[131]

Arkwright's mill at Cromford was extended soon after it was built and further development made it part of an industrial complex fitted into quite a restricted site. A second mill was built in 1776–7. The flow of water to and within the site was now quite complex, with four overshot wheels providing power. The flow of the Cromford Sough was directed to the first mill, and a separate channel from Bonsall Brook brought water to the second new mill.

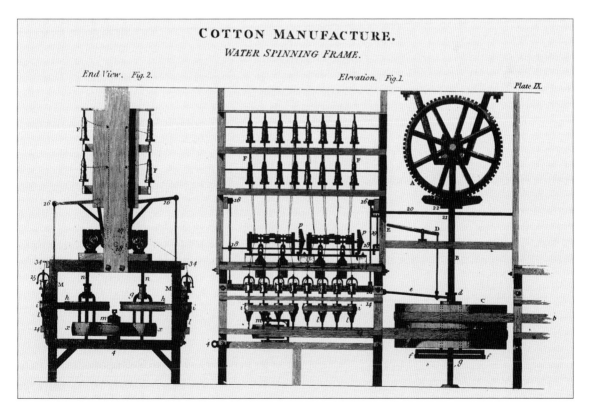

Arkwright's water spinning frame, as illustrated in Rees's Encyclopaedia. *(Museum of English Rural Life)*

Water Power and the Industrial Revolution

At New Mills in Cheshire, a series of mills was built to exploit a short stretch of the River Goyt. Torr Vale Mill was built originally in 1788 as a cotton mill. Extensions involved the installation of a second waterwheel, the housing for both wheels being visible under the empty building. This part of the river has recently had a hydro-electric generator installed on it.

Arkwright's business soon expanded further, and he was building new mills at a number of sites. He had mills at Wirksworth and Bakewell in Derbyshire, and his mill at Shudehill, built in 1780, was the first cotton-spinning mill in Manchester. What became the flagship mill for his enterprise was the new mill at Masson just to the north of Cromford. He bought the land and mill here from Robert Shore and George White in 1780, evidently overcoming whatever doubts he might have had about the Derwent river. Masson Mill, erected in 1783, was a timber-framed building of five storeys in classical style, and powered by a single wheel.

During their partnership, Arkwright and Jedediah Strutt built a mill at Belper, 8 miles downriver from Cromford, completing it in 1776, and another one at Milford. The partnership was dissolved in 1781, with Arkwright retaining the Cromford mills. Strutt kept those at Milford and Belper, where, between 1790 and 1812, he built a series of mills, large in scale both in their buildings and their water engineering.

Arkwright's example was soon followed by many more cotton entrepreneurs, setting up water-powered spinning mills. After Arkwright's patent expired, in 1785, expansion was rapid. Patrick Colquhoun published a census of water-frame cotton mills in 1788, in which he reckoned there were 143 in Great Britain, 124 of them in England.

More recent studies have revised his estimates upwards, to 208 mills in Great Britain. There were concentrations of activity, notably in Derbyshire, following Arkwright's and Strutt's lead, and in Lancashire. The quality of the Pennine streams and the humidity of the area, to which cotton was suited, stimulated rapid growth in the Lancashire cotton industry, with more than forty mills in the south of the county by the late 1780s. Robert Peel's mill built at Bury in 1774 was among the first, followed by a number of others, including Birkacre near Chorley in 1777. When J. Aiken published a survey of the Manchester district in 1795 he remarked upon the clusters of water-powered cotton mills in some localities: twelve at Mottram and five at Royton, for example. Most of the mills used the spinning frame, but a new spinning machine, known as the mule, was developed by Samuel Crompton in the 1770s. Crompton did not take out a patent, and his machine was rapidly adopted. At first it was hand-operated, but water power was applied to it by David Dale at New Lanark in 1792. Thereafter the mule was regularly installed in water-powered mills, becoming the dominant spinning machine within twenty years.[132]

In 1782 Samuel Greg established a new cotton-spinning enterprise at Quarry Bank at Styal, Cheshire, on the River Bollin, where mills had existed since medieval times. His first mill used a wooden waterwheel, but, after extending the mill in 1796, Greg and his partner Peter Ewart reorganized the water power system in 1801. To ensure adequate summer supplies of water, they built a new dam and weir at the same time as putting in two new waterwheels, providing about 20hp of power. This enabled the productive capacity of the mill to increase from 2,425 spindles to 3,452. One of the wheels was replaced with a wheel of wood and iron construction in 1807. In 1818 a third major extension was put in hand so that by 1823 the number of spindles had reached 9,600. To provide the power for this enterprise, a new breast-shot wheel was installed in 1820, 32 feet in diameter by 21 feet wide, and rated at 100hp. It was housed in a chamber underneath the main mill building. This wheel, it was said, then had 'few equals in the country in point of size and efficiency ... its slow and stately revolution seems to be the very embodiment of power and dignity'.

To increase the supply of water to this wheel a new channel was cut, to double the fall to 30 feet, and a tunnel nearly three-quarters of a mile long took the tailrace away. The new wheel replaced the two smaller ones and remained in use until 1904, when a Gilkes turbine of 200hp was installed, supplemented by a 20hp Vortex turbine. Although committed to water power, in 1811, when Greg had 4,000 cotton spindles in his mill, he bought a steam engine of about 10hp to cover for loss of production in drought and to provide auxiliary power. This engine represented only about a quarter of the mill's total power requirements, so Greg presumably expected not to need it too much.[133]

The mechanization of cotton textile manufacture elsewhere in Britain was also accompanied by water power, at least until the 1830s. Between the 1770s and 1790s carding and spinning factories built in Ireland were generally water-powered. As the cotton industry concentrated around Belfast, steam became more prominent. In the 1830s water power provided between 35 and 39 per cent of the industry's requirements, at a time when in the rest of Great Britain 25 per cent of power was derived from water. Decline in the use of water power in Ulster's cotton mills was rapid after 1850, down to 12 per cent of total requirements by 1857.[134]

Quarry Bank Mill, Styal, from the river side. The weaving sheds of the 1830s are nearest the camera, the earlier water-powered spinning mills beyond.

Among the major mills in Scotland were the Catrine Cotton Mills on the River Ayr. They were built in the mid-nineteenth century and extended a number of times. By the early nineteenth century they were operated by James Finlay & Co., who commissioned William Fairbairn in 1824 to reconstruct the wheels and gearing. Fairbairn found the mills worked by four waterwheels, with two 40hp steam engines, brought in as auxiliary power as the reliability of the wheels declined with age. His rebuild involved the replacement of the four wheels with two larger ones, of 50 feet in diameter, having a fall of 48 feet. Each wheel was designed to produce 120hp, enough for the power consumption of the works at the time, but Fairbairn made provision for expansion, with the reservoir and watercourses capable of supporting another two large wheels. The work was very expensive. The wheels, according to Fairbairn's treatise on mills, cost £4,500; £3,000 was spent on work on the headrace, partly in a tunnel, £1,500 on the tailrace, £1,000 on the weir. The costs of land and other works brought the total to £18,000. The new wheels started work in June 1827. 'They have never lost a day since that time,' wrote Sir William fifty years later, 'and they remain even at the present day, probably the most perfect hydraulic machines of the kind in Europe.' Finlay evidently agreed, for, despite the cost at Catrine, he commissioned Fairbairn to install new power at his other works at Deanston. This was another ambitious project, designed to have eight wheels each of 100hp; in the event, four were built between 1827 and 1832. By building a new weir further upstream and a channel three-quarters of a mile long to bring the water down from it, the fall was almost doubled, from 18 feet to 33 feet.[135]

The Industrial Revolution in the cotton industry was founded on the mechanization of spinning. Carding machines to draw out the fibres ready for spinning were an early addition to the mills, but weaving remained a hand process at the end of the eighteenth century. The earliest attempts to mechanize the weaving of the cloth were made during the mid-eighteenth century. Gartside's mill at Manchester, built in 1765, had automated swivel looms powered by a waterwheel. However, there was almost no saving in labour using these machines, and this mill was converted to a water spinning mill. It was not until the early nineteenth century that weaving was successfully mechanized, by which time steam power was becoming a better proposition.[136]

THE WOOLLEN INDUSTRY

Cotton was an upstart newcomer to Britain's manufacturing industry in the eighteenth century, whereas the production of woollen cloth had a centuries-old pedigree. Wool was the first of the manufacturing industries to be a major user of water power, in the process of fulling, one of the finishing stages in the production of cloth. The Industrial Revolution introduced further mechanization to the woollen and worsted industries, but, because spinning machines were introduced later than they were in the cotton industry – in the 1780s – there was always more competition from steam engines to drive the woollen mills.

After fulling, the first of the processes to be mechanized was the preparation of wool for spinning, which involved drawing out the woollen fibres. There are two very similar processes known as 'scribbling' and 'carding', and scribbling was the first to get a machine, during the 1770s. In common with many other machines this could be powered by horse engine, but most were driven by waterwheel. An advertisement of premises for sale at Huddersfield in 1779 was about the earliest reference to a 'scribbling mill'. When the machines were introduced, several fulling mills were either converted to scribbling mills or had scribbling machines added. Armley Mills, near Leeds, and Dewsbury Mills were two established mills in which scribbling machines were installed in the 1780s. Some corn mills were also converted to scribbling mills, including Scothall Mills, near Leeds, on Meanwood Beck.[137]

Conversion of fulling mills was limited, however, because there was often insufficient space to accommodate a scribbling machine. Besides, fulling work was plentiful enough at this time to keep existing mills busy; indeed, there was sufficient demand for some new ones to be opened. Scribbling mills were built most often, therefore, at new sites, of which there were plenty to be had. The Earl of Dartmouth was one of the landowners in the West Riding who actively encouraged industrial growth, and several scribbling mills were built by his estate after 1779. Some of the new mills housed several processes under one roof: Ossett Mills, built on the Calder in 1785, included fulling, scribbling, carding and dyeing. Most scribbling mills were small, requiring modest amounts of power; hence they could be found along the slower-moving reaches of the Aire and Calder near Leeds, as much as on the faster streams of the Pennines. D. T. Jenkins has identified 148 scribbling mills opened in the West Riding before 1800. About eighty of them were powered by water alone, but steam was already being adopted: fifty-four watermills had additional steam power and thirty used steam alone.[138]

Worsted is a form of long-combed woollen yarn, from which fine cloth can be woven, and it was to this branch of the woollen textile industry that spinning machines were first introduced. The first machines were in effect adapted cotton-spinners, able to handle the finer worsted fibre better than the coarser wool. A worsted-spinning mill was built at Dolphinholme, near Lancaster, in 1784, followed by one on the River Wharfe at Addingham in 1787. By 1800 there were fifteen water-powered worsted mills in the West Riding, at such places as Keighley, Shipley and Haworth. Meanwhile, the first steam-powered mills had already been built, including Holme Mill, Bradford, where water power was in short supply.[139]

Water remained important as the woollen industry grew in the eighteenth and nineteenth centuries. Wool had a high price relative to its bulk, which made the cost of transporting it worthwhile. The source of power was a major factor in the location of the industry until steam power concentrated the work in the West Riding.[140] The West Country was one of the areas where water power remained strong as production moved to the factory scale. During the second half of the eighteenth century existing mills were enlarged, fulling mills

Belvedere Mill, Chalford, on the River Frome in Gloucestershire. This nineteenth-century building, on a much older water-power site, was a woollen mill that was converted to silk-milling in the 1850s as the woollen trade declined. Later in the nineteenth century it was used for corn-milling, in which form it continued until the 1920s.

were converted to spinning and new mills were built, most of them sited to take advantage of water power. Among these were many of the largest mills of the south-west region, including the Stanley Mills at Stonehouse, one of the celebrated buildings of the Industrial Revolution, built on the site of a fulling mill and opened in 1813. With five wheels, each able to generate up to 20hp, the mills used water power exclusively until 1822. Ebley Mills, built in 1818, could produce 80hp from their wheels.

Mills for spinning and, later, weaving needed more power. A mill with fulling stocks and a gig mill could work with a wheel of 10hp, sometimes less in a small establishment. A spinning mill needed twice the amount of power, while the large ones demanded much more. Although most of the West Country was well resourced in terms of water power, meeting the demands of an expanding industry was not straightforward. The gentle gradients on most rivers meant that millwrights usually settled for undershot or low breast-shot wheels. The use of more than one wheel, as at Stanley Mills, was one means of ensuring adequate power. Water supply was ensured by the construction of reservoirs. At Stanley the reservoir extended ultimately to 15 acres, held by a dam 30 feet high and 150 yards long. At Ebley a large meadow was flooded, and all over the region ponds and leats were constructed to feed the waterwheels. Nevertheless, growth of the woollen industry did put great pressure on available sites along the rivers of Gloucestershire: there were twenty-three mills in 4 miles in the Stroud valley, for example. By the 1820s many rivers were over-crowded; summer drought and inadequate flow to mills downstream were common problems. This was a peak time for the industry in this area before it succumbed to a fluctuating, downward trend.[141]

Steam power was being installed in the rapidly growing mills of Yorkshire, the prime centre of the woollen industry, from the 1790s onwards, and this was a fairly general picture within two decades. It was an earlier and more rapid transition than that experienced by the Lancashire cotton industry. Bradford, the 'capital' of the industry, and the district south of Leeds, were badly off for water supplies, and the proximity of the coal fields also discouraged the use of water power. Most of the worsted mills were located in this district, and the factory inspectors' returns for 1850 showed only 13 per cent of the power used in the worsted industry coming from water. Higher up in the Pennine valleys, around Halifax, Saddleworth, Skipton, Keighley and Otley,

Inside the spinning shed of a Welsh woollen mill powered by water in the 1930s. (Museum of English Rural Life)

water power was more common. It was in this area that one of the first of the mechanized worsted-spinning mills powered by water was opened, at Low Mill, Addingham, near Skipton, in 1787. The Pennine mills were generally smaller than the steam-powered mills further south, but the cheapness of the power meant that they kept going with water until later in the nineteenth century.[142]

The woollen industry of the West Country retained a stronger base of water power. In 1838, 66 per cent of the power used in Gloucestershire and 59 per cent in Somerset came from water. The great exception in the use of water was Trowbridge, where the streams represented an inadequate source of power. Steam was introduced from an early date to the mills at Trowbridge, with the result that by 1838 the contribution of water to the total power consumption in Wiltshire was down to 32 per cent. In Devon, all the mills were powered by water.

In Scotland, dependence on water power was high, for most of the woollen mills were in areas remote from coal supplies, such as the border country.[143]

In the valleys of west Wales the manufacture of flannel expanded during the first half of the nineteenth century, with water as its power. Llanidloes, where the rivers Severn and Clywedog meet, became one of the principal towns for the Welsh woollen industry during the Middle Ages. Trade was growing during the early nineteenth century and by 1838 there were nine substantial mills and a number of smaller ones on the two rivers. However, the factors that had aided the growth of the industry – the water power, and the small, locally based factories – later became a hindrance. The Welsh industry was always short of capital and unable to expand much because there was a limit to what could be achieved with the water power, while most were distant from supplies of coal. It proved impossible to compete effectively with the expanding woollen industry of Yorkshire, and the Welsh industry declined rapidly after 1880. Although, in 1895, mills making flannel, shirts and blankets could still be counted in their hundreds, most of them were still small and water-powered.[144]

THE LINEN INDUSTRY

Water power lay behind the growth of the linen industry in Ulster. According to one of the province's historians, 'It is not too much to say that it was largely as a result of the availability of considerable water power potential that an area without the basic ingredients of industrial strength in the technological revolution of the period 1730–1880, namely coal and iron, was able to advance to a position of worldwide importance.'[145] From modest beginnings in the early eighteenth century, the linen industry of Northern Ireland had grown to more than one million spindles by the late nineteenth century. Mechanization, and the need for water power, was first applied to the finishing processes of linen manufacture, beetling and washing or bleaching. Beetling involved pounding the finished cloth to bind the weave together, improving its softness and impermeability. It was a similar process to the fulling of woollen cloth. In the early nineteenth century, beetling was also applied to the coarse yarn. The hand process used heavy wooden mallets ('beetles'); the machine, introduced in Ireland around 1725 and in Scotland in the 1730s, used a series of hammers (almost always of beech).

The two processes, beetling and bleaching, were usually run together at works known as 'bleach greens'. Until the arrival of mains water, bleachers needed a site with copious water supplies, usually outside the towns; if there was enough water for the bleaching, what was needed for the beetling mill wheel was almost a free ancillary. As the linen industry grew during the eighteenth century, so did the number of bleach greens. There were about 200 of them by 1760, and 357 in 1787, but the numbers reduced during the following century as the industry restructured and contracted. Water power remained important in these finishing processes, since the extra speed of a steam engine was not essential. Most bleach greens had two or more waterwheels. Even in 1930, a quarter of the power other than electricity used in flax finishing works was provided by water.[146]

The early mechanization of the linen finishing processes bestowed wealth and influence upon the owners of bleach greens, so that, when mechanization reached spinning and weaving, much of the capital came from bleachers. The first trials with mechanical spinning took place at the end of the eighteenth century, and the first power weaving plant was opened in Cork in 1814, with others following in County Monaghan in 1825. All these early factories were water-powered. By the time mechanization had become more widespread, in spinning from the 1830s and weaving from the 1850s, steam was more likely to be the choice of power, certainly for the large urban factories. Outside the towns, waterwheels continued to make a contribution, and as much 15.8 per cent of the power used in Ulster's linen-weaving factories in 1875 came from water.[147]

Before the linen was manufactured the useful fibre was extracted from the flax by the process of 'scutching', which entailed passing it through rollers. A water-powered scutching machine was introduced in the early eighteenth century. As the linen industry expanded, so did the number of scutching mills – 630 in Ulster in 1830, and a peak of 1,420 by 1868. Most were still powered by water, and several were located in converted corn mills. Scutching was mainly a small-scale seasonal process, carried out for six to eight months of the year. Mills were mostly located in rural areas, and the business was regarded as ancillary to farming rather than as part of textile manufacturing. This helps to explain the continued adherence to water power, although numbers of mills declined with the increasing use of prepared flax imported from Russia and Belgium. Of the 600 scutching mills at work in Ulster in 1914, about three-quarters were water-driven, although in many turbines had replaced the vertical wheel. The last of these stopped work in 1958.[148]

THE PROS AND CONS OF WATER POWER FOR TEXTILES

Between the 1760s and the 1780s the factory masters had no real choice other than to use water power, but it did have an appeal in its own right for the textile mills. First, the waterwheel moved steadily, with little fluctuation in its motion, and this was further enhanced with the introduction of the 'governor'; transmitted to the machinery, steady motion reduced the number of breakages of the yarn. This benefit was so highly valued that many mill owners preferred water power long after the steam engine had become just as practical a proposition. Water power was thus still being developed in the industry into the 1830s, even though the use of steam had been advancing steadily since about 1800.[149]

The factory owners wanted a steady flow of production: theirs was not a seasonal occupation and they needed the machines in constant use. One major worry for them was a failure of water supply, whether from extremely dry seasons or from flood. Maintaining a steady rate of production was so important that at Belper, and at some other mills, a clock worked off the wheel to record 'mill time' – the amount of work done in the hours of working. Time lost through shortage of water had to be made up.[150] Drought and flood were hazards from the beginnings of the Industrial Revolution. The Strutt mills at Belper had flood wheels, which were brought into use only when the river was at maximum flow, but better wheels and bigger reservoirs were never the guaranteed solution. Textile mill owners would seek a location that offered a year-round water supply, or as near as possible, as Arkwright had done at Cromford. The Pennine hills, with their high annual rainfall, clearly had an attraction. On the peaks from which the streams drew their sources, as much as 70 inches a year might be recorded; further down the hills, where most mills were built, over 40 inches was usual for many places. Manchester, with 30 inches of annual rainfall, was one of the drier spots, and was not a popular place for water-powered mills. Bolton also failed to attract many mills before the advent of steam power because the number of sites in this area with a plentiful, constant flow of water and a reasonable fall was limited. Stockport, conversely, became a major mill town at a point where the gradient of the Mersey became steeper. Between 1736, when water power was first used for silk manufacture, and 1800, the gorge was filled with water-powered mills.[151]

THE INDUSTRIAL REVOLUTION BEYOND TEXTILES

The effects of the Industrial Revolution ranged far beyond the textile industries – to iron, engineering, pottery, paper-making, mining, and many more industries. Water power was involved in many of these, as provider of power, and sometimes as stimulus for development.

Advances in the productivity of iron manufacture, with the introduction of coke-fired blast furnaces by Abraham Darby in 1709, was a third major strand in eighteenth-century industrialization alongside the cotton and wool textile industries. Water power, which for some 300 years had been employed in this industry, continued to have a role, for the wheel was needed for working the bellows in the early coke-fired blast furnaces. However, in Coalbrookdale in Shropshire, where Abraham Darby was based, the lack of supply of water soon became a limiting factor in the expansion of production. In the 1720s production was seasonal, starting in September/October as reservoirs were replenished, working up to a peak of production between December and February, and slowing down during spring and early summer. Summer water shortages meant that production was reduced and usually stopped completely for a few weeks. A year of particularly low rainfall in 1735 led to the use of horse-worked pumps to recycle the water from the tailrace back up to the reservoirs. A Newcomen steam engine was bought in 1742 to replace the horses in this pumping operation, and this enabled production of iron to increase during the second half of the century.[152]

Similar difficulties were experienced by other ironworks, such as Calcutts (opened around 1770) and Benthall, built in the Severn valley in the wake of Coalbrookdale's success. They, too, were using steam pumps in the late eighteenth century to recycle water to the wheels that drove the bellows, boring mills, rolling mills and hammers. In 1800, Thomas Telford supported proposals for controlling the flow of the Severn on the grounds that it would overcome the shortages of supply for the water-wheels at the ironworks. By this time, however, the Shropshire industry was losing its competitiveness against new blast furnaces opening elsewhere, which used steam engines for most of their power requirements.[153]

The metal-using industries were common users of water power. Many had modest requirements, for hammering, cutting, grinding and sharpening, and

One of the overshot wheels at Finch Foundry fed by a launder coming in from the left. (Museum of English Rural Life)

Water Power and the Industrial Revolution

Forge hammers at Finch Foundry, Sticklepath, Devon, 1966. The cam wheel to trip the hammers is visible behind them. (Museum of English Rural Life)

water power served very well for all of these. The nail-makers, and needle- and pin-makers to the south of Birmingham were all users of water power. The area round Redditch, for example, where mills on the River Arrow were grinding needles, was one local concentration of such activity, while edge tools had been the speciality of Belbroughton, in Worcestershire, since the fifteenth century. The independent craftsmen of the area were consolidated into a single business during the nineteenth century. In the West Country, edge tools were what James Fussell made at the ironworks he leased near Frome in 1744. Expansion by the end of the century had added further works and a wide range of production, of all types of sickle, billhook and scythe, together with spades and other tools. William Finch converted a woollen mill at Sticklepath, Devon, to iron working in 1814, where again edge tools were the main products. The old manorial corn mills were later taken over and turned into a finishing works. Spade mills were common in northern Ireland in the eighteenth and nineteenth centuries: there were at least twenty-one in the province in the 1830s.[154]

The expansion of the tool-making and cutlery trades of Sheffield during the eighteenth century created another concentration of water-powered industry, with large numbers of works, each quite small, drawing power from the rivers. In the 1770s there were 161 works on the five rivers of Sheffield, using water power in the various metalworking trades – cutlery, tools, forgings. The cutlery trade alone was said to employ no fewer than 133 wheels, driving bellows for hearths, hammers, cutting tools and grindstones. Abbeydale, on the River Sheaf south of the city, was one of the largest of these works. It is first recorded in 1714 but may be much older – certainly, there had been industrial activity in the area since the thirteenth century. The dam was enlarged in 1770 and a new forge constructed in 1785. By the end of the nineteenth century several forges and grinding shops had been added, the power coming from two wheels, one high breast-shot and one overshot. Production was mainly of edge tools, but other tools were also made.[155]

Matthew Boulton started out in business with an iron-rolling mill. He leased Sarehole Mill in 1759, converting it from a corn mill to a rolling mill, but the water power available was insufficient for the work. Boulton moved to Soho, the last undeveloped water-power site left in the vicinity of Birmingham, leaving Sarehole to return to grinding corn. Soho also had problems with water supply, especially in summer, so Boulton turned to steam pumps to feed his reservoir. A Savery engine at first was followed by one of James Watt's, and from this deal their partnership as steam-engine builders emerged. Water power continued to be used at Soho for some time: John Rennie installed a new waterwheel in 1784, just as James Watt was successfully devising his rotary drive.[156]

The growth of the potteries in the eighteenth century stimulated a demand for ground flint and stone. White earthenware, in particular, one of the main innovations of the time, required crushed flint and stone to mix in with china clay. Watermills, many converted from other uses, were turned to this grinding work. The flint was ground down to a white slurry in a process that could take twelve hours. There were some local concentrations of these mills. One of the principal areas was in Staffordshire, along the Trent and its tributaries, such as the Churnet, conveniently placed to supply the earthenware factories. At least twenty-seven mills are known in this area, around Stone, Cheddleton and Burton-upon-Trent. The mills were small in scale, producing perhaps no more than one ton a week, but demand, and the price, were so high that the miller could make a living and pay the high rents demanded by the landlord. A second major concentration of water-powered stone-grinding mills existed in Cornwall. Elsewhere in the country, flint-grinding mills served local potters; seven were located along the Ouseburn, supplying Newcastle-upon-Tyne. Bones and other material for glazes were subjected to the same processes.

As early as 1782 Josiah Wedgwood invested in a Boulton & Watt steam engine to crush flint, and steam mills with ten times the output of the

Sarehole Mill, south of Birmingham, was where Matthew Boulton first set up business rolling iron. It subsequently returned to grinding corn, in which form it is preserved.

watermills were soon having an impact on the trade. Even so, there was sufficient demand for some to continue at work until the early 1960s. Water power for stone-cutting was sometimes found too, such as at Ashford in Derbyshire, for the cutting of 'black marble'.[157]

Although the technology of powered saws was long established, it was not until the increased demand for timber in the eighteenth and nineteenth centuries, and the need to handle it in larger quantities, that it was taken up to any extent in Britain. Wind and water were both used to drive sawmills in the eighteenth century. There were not many powered mills and they were confined to the larger timber yards, such as the one on the River Itchen owned by William Taylor. Taylor had naval contracts and used water power as an aid to greater accuracy in cutting ships' blocks. From the late eighteenth century onwards, the number of sawmills for timber increased, both commercial enterprises and on landed estates. There were about sixty sawmills in Scotland by 1830, working local and imported timber. Among the English estates, Gunton in Norfolk had a sawmill driven by an overshot waterwheel installed in the 1820s, and Fountains Abbey estate converted its old corn mill for timber work.[158] However, by the middle of the century most estates were opting for steam power: a small portable engine and a sawbench could conveniently be used, either in the estate yard or taken out to the stands of timber.

Mills powered by water and wind were also used for the grinding of snuff. The taking of snuff became fashionable during the eighteenth century, prompting production in Britain to add to imported supplies. Many of the mills were near Bristol, where the tobacco was imported, and there were also watermills in Wiltshire engaged in this trade. Other mills were operated by the Wilson family in Sheffield.[159]

Mining and quarrying continued their use of water power, mainly for pumping water, but also for such work as drawing trucks out of the mines. Water power had been introduced to the lead mines of the Pennines during the late seventeenth century. Expansion of the industry from the 1780s onwards led to greatly increased use of water power, encouraged by the improvements in the power and efficiency of waterwheels. Water power continued to be installed in the nineteenth century in preference to steam, sometimes even replacing steam engines, as happened at Derwent Mines, County Durham. Waterwheels were installed in large numbers – there were at least seventeen at the mines on Grassington Moor, Yorkshire – with many of those for pumping being placed underground. As well as waterwheels, hydraulic pressure engines were used for pumping and lifting. Water power served other purposes at the mines, too, such as crushing and sieving the ores, and in the 1870s, driving an air compressor for a rock drill.[160] One smaller quarrying activity produced limestone for agriculture and lime pits at Nurcott and Newland in Exmoor had waterwheels for pumping and transport in the mid-nineteenth century.[161]

Small-scale wood crafts and turnery used water

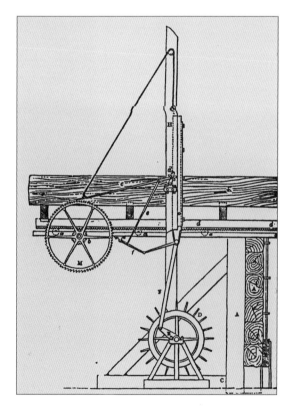

A sawmill built for Chatham dockyard early in the nineteenth century. From Joseph Glynn, Rudimentary Treatise on the Power of Water.

power in varying degrees, and the turning of bobbins for the textile industry became a significant activity in the eighteenth and nineteenth centuries. The work was handled mainly by small water-powered mills in the southern Lake District and West Riding, built on the basis of abundant supplies of coppiced wood and the availability of water power. The village of Staveley in Westmorland, at the meeting point of three rivers and on the main turnpike road, became the main local centre for the bobbin trade. In the eighteenth century Staveley already had a market and by 1831 there was a cattle fair as well. By the middle of the nineteenth century there were nearly fifty mills at work in the Lake District. Most were fairly small, but a few, such as the one built at Stott Park, Windermere, in 1835, with a 35-foot waterwheel, had substantial power-generating capacity.

The bobbin mills of the Lake District were a major source of employment in this rural area, especially of boys and young men, and helped to stem the flow of migration to the towns. The peak of the industry was reached in about 1860, by which time the number of mills had reached sixty to seventy. From the 1870s a decline set in, prompted by a number of circumstances: some of the bigger customers in Lancashire and Yorkshire had started to turn their own bobbins; supplies of local wood were becoming less plentiful; and competition from Scandinavia, in the supply both of timber and of bobbins, was increasing. The number of mills rapidly diminished, but Lakeland bobbin-making was not entirely finished. Several water-powered wood turneries were still active in the late nineteenth and early twentieth centuries, often with new turbines replacing the vertical wheels.[162]

There were water-powered wood-turneries in other parts of the country, making domestic and industrial goods, such as the brush and broom works at Bradiford, north Devon. This firm closed in 1935 when a fire destroyed the works, while others succumbed to competition from imported products.[163]

There remained a host of industries and smaller trades using water power in one way or another. There was hardly an aspect, it seems, of economic life in which water power was not involved. In the distilling and brewing trades, waterwheels were used for grinding the malt. Bushmills distillery in Northern Ireland was still using water power in this way in the twentieth century, updated with a turbine installed in 1928.

TRANSPORT

In order to maintain water supplies, canal companies used pumps as well as feeder channels from rivers. Most were steam-powered, but a few waterwheels were used, including one at Welshpool on the Montgomeryshire Canal and another at Mellingriffith on the Glamorganshire Canal built in 1807. At Claverton, near Bath, on the Kennet & Avon, John Rennie built a pump driven by a breast-shot wheel that was 24 feet wide in 1813.[164] A waterwheel was installed in 1803 to drain a canal tunnel at Morwell Down, near Tavistock in Devon. The wheel, which was 40 feet in diameter, drained water from the tunnel, which was one and a half miles long. Pumps driven by waterwheel aided the ventilation of the tunnel.[165]

One of the means by which canals used to carry boats up and down hills was the inclined plane, and waterwheels were sometimes used to operate them. The Bude Canal in Cornwall had five water-powered inclined planes along its fairly short length in the 1820s. There was another example at Morwellham on the River Tamar, on which a waterwheel drove the winding gear to lift boats 240 feet. Canal-side warehouses sometimes used water power for lifting gear. James Brindley installed a wheel for lifting coal in a warehouse on the Bridgewater Canal in the 1760s, and one from the 1820s in some warehouses on the Rochdale Canal in Manchester.[166] Before steam and locomotives became the dominant form of motive power, water was used occasionally for traction on railways. When it was built in 1836, the Ffestiniog Railway in north Wales did not use locomotives. Most of the traction was by horse and gravity, but one of the rope-worked inclines at Moelwyn used a waterwheel supplied from a dammed stream.[167]

8 Developing the Technology of Water Power

Many of the new water-powered factories at the beginning of the Industrial Revolution required no greater power than the corn mills, fulling mills and iron forges of earlier generations. Arkwright's first mill wheel at Cromford, for example, needed no more than about 12hp. The demand for power increased rapidly, however. When the West Mills at Belper were built, 25 years after Arkwright's, the wheels could produce about 80hp. By the 1830s the Strutts had eleven wheels for their Belper mills, six of them reserved for high water. The firm estimated their output then as 200–300hp. In the nineteenth century the paper mills at Pickwash below Duffield, Derbyshire, had the right to take as much as 800hp from the Derwent, while Pegg & Ellan Jones, paint manufacturers of Derby, established in 1820, used water power of up to 300hp.[168]

The demands of the new industries of the eighteenth century prompted major advances in the technology of water power. Textile factories wanted more power; gearing and shafts had to be designed to take the power throughout a large mill building. The work of engineers and millwrights such as John Smeaton, William Fairbairn, Thomas Hewes and Joseph Glynn contributed to the design of improved mill wheels and gearing, while the firms established by many of these men set the manufacturing standards.

JOHN SMEATON AND THE EFFICIENCY OF WATERWHEELS

John Smeaton had a dominant influence on the technological development of the watermill in eighteenth-century Britain. His experimental work on waterwheels demonstrated that the overshot wheel was more efficient than the undershot, and this became part of the canon of millwrighting knowledge. His work as an engineer helped transfer his experimental results into everyday practice, and he was also instrumental in introducing iron as a common, necessary material for the mill's construction.

Smeaton was not alone in seeking to improve the efficiency of the waterwheel. Mathematical principles had been applied to its operation by European scientists going back at least to Galileo. In the first half of the eighteenth century a number of scientists, working mainly in France, made calculations relating to the force of water and the velocity needed on the blades of a waterwheel to achieve maximum power. Antoine Parent was one of the most prominent, and his work became known in Britain through the writing of engineers such as J.T. Desaguliers. Parent made a calculation of the amount of power that a given quantity and given fall of water striking the blades of an undershot wheel would produce. He went on to calculate how much of that power was turned into actual mechanical power transmitted through the wheel. He reckoned that only about 15 per cent of the power from an undershot wheel was usable. This was a percentage of efficiency that entered into common parlance amongst millwrights and engineers. Parent further proposed that the waterwheel worked at its best with water applied to about one-sixth of its circumference. Other investigators, such as Henri Pitot, worked from Parent's conclusions to propose designs of wheels with blades distributed in such a way that would make the best use of the water's power.

By the middle of the century, confidence in these

results had diminished. Experimentation by Antoine de Parcieux indicated that the overshot wheel worked better than the undershot, and it was into this area of research that John Smeaton entered.[169]

Smeaton advanced his investigations into the effectiveness of the waterwheel by experiment rather than mathematical calculation, and by considering the overshot wheel as well as the undershot. He had started his career in his father's law firm but, finding that he had a greater aptitude for engineering, had turned to making scientific instruments. His broad interests in engineering led him to work on mills, first at Halton, Lancashire, in 1753, then at Wakefield in the following year. Perhaps inspired by this work, in 1752–53 he began to conduct experiments on the power produced by waterwheels and windmill sails, using working models. After further research, he was ready to present his conclusions to the Royal Society in 1759, in a paper entitled 'An Experimental Enquiry concerning the Natural Powers of Water and Wind to turn Mills and other Machines depending on a Circular Motion'. Smeaton was awarded the Society's prestigious Copley Medal for this paper, which caused quite a stir and was later published in book form.

Smeaton's reputation as an engineer was firmly established as a result of this work. The commissions for watermills increased significantly, including work on waterwheels for iron forges, textile mills, water pumping and other applications. One was a large wheel for the corn mill at the naval victualling yard at Deptford. As there was no natural fall of water to speak of, a Newcomen steam engine was used to pump water to the reservoir above the wheel. From working on watermills, he went on to become the nation's leading consulting engineer, building the Eddystone Lighthouse, as well as docks, canals and bridges.

Much of the excitement about Smeaton's work on waterwheels had been caused by his principal findings on their efficiency. He demonstrated, first, that undershot wheels were, at 30 per cent, more efficient than the calculations of Parent and others had allowed. Furthermore, overshot wheels were twice as efficient, at 66 per cent efficiency. His work also showed that there was a difference in effect between the impulse machine of the undershot wheel and the gravity machine that was the overshot, to the latter's advantage. There had been a strongly held view that impulse and gravity wheels were equally effective, that power output resulted only from the mass and velocity of the water. Smeaton found that the overshot wheel worked more efficiently at very low

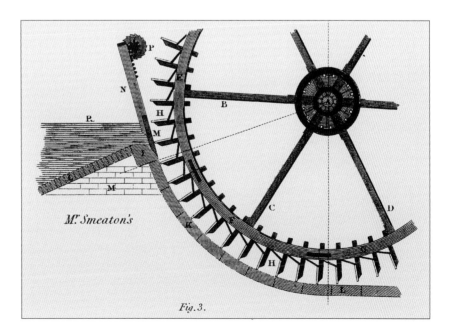

A breast-shot wheel designed by John Smeaton, illustrated in Rees's Cyclopaedia. *(Museum of English Rural Life)*

speed – the optimum speed at the circumference of the wheel was 3½ feet per second. Later work, notably by the Franklin Institute in the United States, showed that there was a marginal loss of efficiency from higher speeds. In practice, most wheels did turn faster, with 6 feet per second being common.

Smeaton's work was very timely, for the new factory owners were starting to look for greater power, and he offered them proof that more could be gained from a river by using more efficient wheels. John Farey, an engineer who wrote about the development of steam engines in the early nineteenth century, believed that the impact of Smeaton's work was sufficient to delay the application of steam power. It certainly resulted in great enthusiasm for overshot wheels among milling engineers in Britain. At the beginning of the eighteenth century, the undershot wheel was almost certainly still the most common type. By the end of the century, received wisdom was that the undershot wheel was somewhat anachronistic and the overshot was a better proposition.[170]

As things worked out, however, the predominant waterwheel in the British Industrial Revolution was the breast-shot type – neglected for centuries, it suddenly became a favourite. It was the type of wheel that provided power for most of the textile mills, and a good many of the corn mills of eastern England. Even though he would use overshot wheels where possible, about 70 per cent of the wheels supplied by John Smeaton were of the breast-shot type. According to Rees, in his *Cyclopaedia*, 'It was one of the continual occupations of Mr Smeaton, during 40 years, to improve the old watermills by substituting breast wheels for undershot.' William Fairbairn also acknowledged that the breast-shot wheel had 'taken precedence' over the overshot wheel.[171] Partly this was down to circumstance. Often, conditions would not allow an overshot wheel to be installed. In mining districts, high aqueducts were often built to feed the wheel, which was feasible in hilly districts, but more difficult on lower rivers. Rivers crowded with textile mills, such as the Derwent and the Bollin, had insufficient fall along most of their course for overshot wheels. Much the same was true of many major corn-milling rivers in southern England, while the low-lying terrain of East Anglia gave no opportunity for sufficient fall, not even with the most extensive of water channels. In districts where the fall was no greater than about 4½ feet the undershot wheel remained the most common, and the next best thing as far as many milling engineers were concerned was a breast-shot wheel.[172]

This may make the breast-shot wheel seem like a

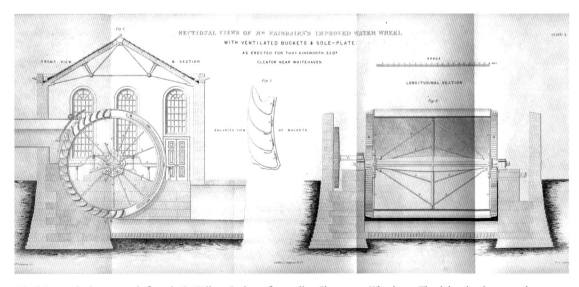

A high-breast wheel, constructed of iron by Sir William Fairbairn for a mill at Cleator, near Whitehaven. The sliding hatch gate can be seen, as can the improved ventilated buckets. (Museum of English Rural Life)

Curved floats fitted to the waterwheel at Terwick Mill, Rogate, Sussex, which was still working when it was photographed in 1962. (Museum of English Rural Life)

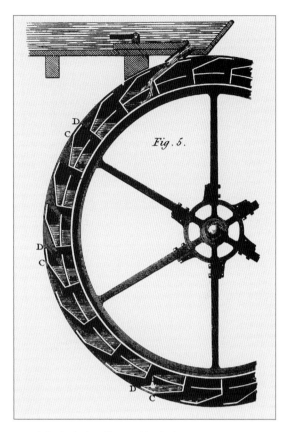

A pitch-back wheel as illustrated by Abraham Rees.

compromise – and for many millwrights it probably was, at least initially – but the work of Smeaton and others turned it into a very efficient engine, effectively possessing the advantages of the other types of wheel, but with few of the disadvantages. It became something of a commonplace amongst engineers that a breast-shot wheel could actually outshine an overshot for efficiency.[173] With a well-constructed brick or stone breast (the part of the wheel that gave the type its name) to feed the water to the wheel, and an iron construction, there was often little to choose between this and the overshot wheel. Breast-shot wheels were likely to cost less to install, mainly because the construction of conduits to feed the wheel was usually less complicated. The conduits bringing water at a height needed for an overshot wheel could be, according to Fairbairn, 'as difficult of construction and as expensive as the weir. In several large works with which I have been connected, the cost of conduits has extended to many thousands of pounds, as at the Catrine works in Ayrshire, or the Deanston in Perthshire.' Fairbairn argued that the breast-shot wheel would cope with the variable head of most rivers without any loss of efficiency. It could thus allow the mill owner flexibility in drawing the water from his reservoir, and keep going for longer at times of low rainfall. Furthermore, he said, a breast-shot wheel of large diameter would deal with the increase of back water at time of flood – the clockwise motion of an overshot wheel could actually scoop floodwater back into the wheel. The capacity to cope with back water was a reason for the continued use of undershot or low breast-shot wheels in areas of low gradient, such as the West Country.[174]

There was little loss of power in using a breast-shot wheel and the general reckoning was that it was about 55–60 per cent efficient. Fairbairn reckoned that a high breast-shot wheel constructed of iron could be as much as 75 per cent efficient, and this was the most power that could be obtained from a water machine. Not everyone agreed: others thought breast-shot wheels were generally about

Developing the Technology of Water Power

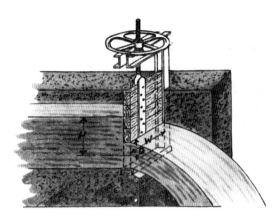

The common form of lifting sluice gate, which allows the water to flow underneath. (Museum of English Rural Life)

two-thirds as efficient as an overshot. Smeaton's work had shown that the breast-shot wheel could be made more effective by setting the paddles at an angle or even replacing them with buckets, as in an overshot wheel. The wheel was thus propelled more by gravity than by impulse. Wheels of this type, with the water dropped in at the highest possible point and gravity pushing the wheel in an anti-clockwise direction, were known as 'pitch-back wheels'. Since they needed a large diameter if they were to work well, their development came from the early nineteenth century onwards, when iron construction had become usual.[175]

In the 1780s, John Rennie introduced the sliding hatch, or overflow sluice gate, inspired, it was said,

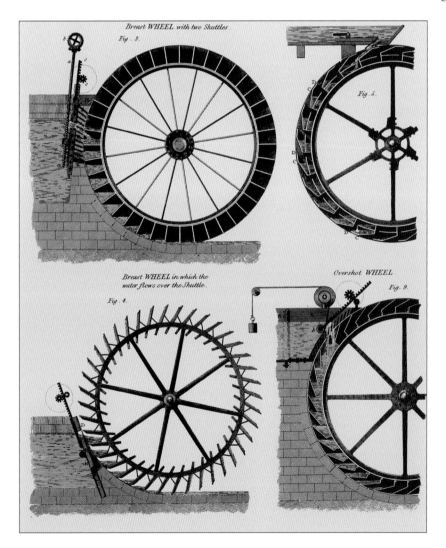

The illustrations in Rees's Cyclopaedia *show the workings of the improved sluice gates for waterwheels. The low breast-shot wheel at bottom left – a representation of a wheel built for the Royal Armoury Mills, Enfield Lock – is shown with Rennie's hatch, which is opened by lowering the gate, to allow water to flow over the top. At top left is a development of the principle with two openings to the gate. A counter-balanced gate for a pitch-back wheel is on the bottom right. (Museum of English Rural Life)*

by observing the way water spilled over the mill dam near his home in Houston, Lothian. Whereas sluice gates previously had been raised to let the water flow underneath to reach the wheel, the sliding hatch worked in the opposite way, by lowering the gate and allowing water to flow over the top. It was easier to operate the gate in this way, especially when closing it against the force of the stream, and this allowed a more minute adjustment than the lifting gate. The operator thus had more control over the flow, and the wheel was better able to cope with varying levels of water. Closer regulation of the flow of water helped ensure that the breast-shot wheel was adopted in preference to the undershot wheel wherever economy of water was a serious consideration. Rennie later adapted his sliding gate with the addition of a double hatch, which allowed even more accurate regulation. This was something that suited textile millers particularly, with their need for a steady speed of work.

Improving the efficiency of the floats and buckets was another concern of the engineers. Water could be lost, and with it power, in two ways. As the full bucket of an overshot or breast-shot wheel descended, water might spill out well before reaching the bottom of the revolution. There was discussion of the merits of partially filling the buckets, making the mouth of the bucket more curvilinear, and fitting the masonry in the breast of the wheel-pit as tight as possible. All these options might reduce spillage, but none was a perfect answer. The second cause of power wastage might be dealt with more successfully. As the water filled the buckets of overshot and breast wheels, air was forced out. If the air could not escape quickly enough, water would spill out, with consequent loss of power, sometimes by a considerable amount, with spray occasionally rising as high as 6 feet. Drilling holes in the buckets was an effective remedy – except that some water could also escape through the holes.

William Fairbairn applied himself to the problem from the 1820s onwards. His eventual solution was to insert small air vents at the top of each bucket, leading the escaping air into a passage on the inner rim of the wheel. He presented a paper about his

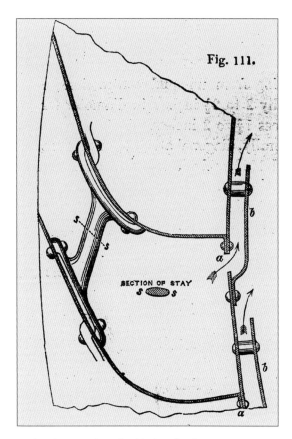

Fairbairn's improved ventilated buckets, from his treatise on water power. (Museum of English Rural Life)

design to the Institution of Civil Engineers in 1849, claiming a gain in efficiency of 25 per cent from the improvement. By that time he had built a number of wheels to his new design, the first at Handforth, Cheshire, in 1828. Many other orders followed.[176]

THE PONCELET WHEEL

In France in the nineteenth century there was a strong desire to increase the efficiency of watermills. France was short of coal and water power was seen by many as the best way to make progress with industrial development. As a result, while in Britain interest declined somewhat after Smeaton, work continued in France on improvements to water technology. One major development led to the modern water turbine, while another was an improved

Developing the Technology of Water Power

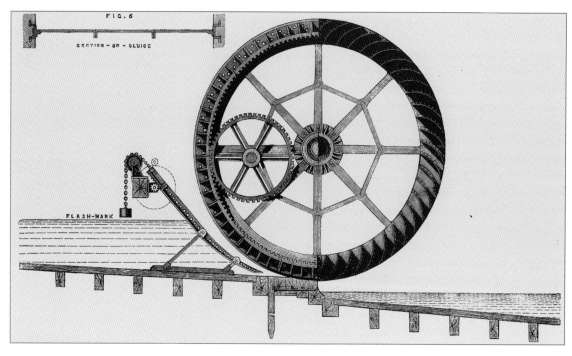

The Poncelet waterwheel installed at Buscot Park, Oxfordshire, in 1866, where it was used to pump water. The wheel of 16 feet diameter had fifty-two blades, operating on a fall of only 3 feet 6 inches. The curved sluice gate was intended to feed the water more evenly on to the wheel. The illustration is from The Engineer. *(Museum of English Rural Life)*

undershot wheel designed by Jean Victor Poncelet, who published its details in 1825. Poncelet worked from the mathematical principles established by the scientists of the eighteenth century to design a wheel that had curved blades in place of plain floats. The size of the blades was carefully calculated, and their number could be considerably more than the floats on an undershot wheel, but, because the wheel did not support the weight of the water, the whole could be of quite light construction. Side panels were usually fitted, so that the water was enclosed as much as possible, in order to maximize the force on the wheel. The water needed to be delivered as close to the wheel as possible, which usually necessitated some construction work on the water channel, to create a downward gradient and force the stream into a narrow passage. The wheel was thus far removed from the ordinary undershot wheel, acting more like the impulse turbines that were being developed at the same time. Theoretically, these wheels could be as much as 80 per cent efficient, but in working conditions most were approximately 60 per cent efficient. It was a great improvement on the original undershot wheel.

Poncelet's wheel was not often used in Britain. By the time it was introduced, steam power was a more serious competitor. The precision of its design made it expensive to produce and to install, and this was a sufficient disincentive for most mill owners. As a result, when a Poncelet wheel was installed it tended to attract attention. In 1866, *The Engineer* included a report on one such installation, when Robert Campbell, the owner of Buscot Park estate in Oxfordshire, decided to employ a Poncelet wheel to work water pumps for land irrigation. It was 16 feet in diameter, working from a head of water of 3 feet 6 inches.[177] Campbell had plans to put in further wheels of this type, but when the time came he had moved on to turbines. Poncelet wheels were put in at Exwick Mill, Devon, in the 1880s (also featured in *The Engineer*), one or two were in use on the upper Bann river, northern Ireland, and Poncelet-type wheels were installed in the tide mills at Eling, Hampshire, and Three Mills. Most British milling engineers looked for a cheaper alternative. Sir William Fairbairn was of the view that, while the Poncelet wheel had the advantage where there was a fall of up to 6 feet, above that the low breast-shot wheel was equally good for most purposes. In general, millwrights were happy to take the idea of curved floats and fit them to ordinary undershot and breast-shot wheels. That in itself effected a gain in power and efficiency.[178]

At the beginning of the twentieth century, the firm of R. G. Morton at Errol, Perthshire, made some alterations to the Poncelet design, enclosing it to make it into a turbine wheel. This, they claimed, improved on the efficiency of the Poncelet wheel. They sold one of these wheels, made of aluminium and working from a head of nearly 200 feet, to Invermay House in Scotland to generate electricity.[179]

SIZE AND HORSEPOWER

Wheels of large diameter became more common from the second half of the eighteenth century onwards. They depended on two other developments of that time: the use of rim gears to transmit power from the wheel instead of through the wheelshaft, and the use of lightweight iron for the spokes. Wooden waterwheels of compass-arm arrangement had been built with a simple geometrical arrangement of between four and eight solid spokes. Iron wheels initially followed the same pattern, using cast-iron spokes, which had to be of a similar size to the wooden ones. Wrought iron, however, had fewer impurities than cast iron, and spokes could be made stronger but lighter. Wheels could now be built with a larger number of flat arms of wrought iron of smaller cross-section, enabling greater diameters to be achieved. The development of the suspension wheel, tied together by many light wrought-iron tension rods, took this principle a step further. This produced wheels combining strength with light weight, allowing greater flexibility in design and construction.

Wheels of a much larger diameter could be built using this construction. By the mid-eighteenth century, millwrights using solid-spoked construc-

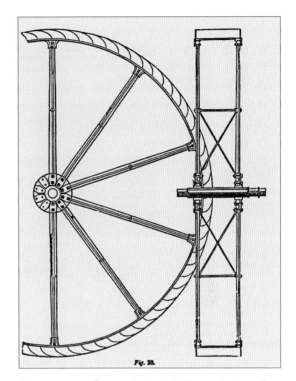

The construction of a waterwheel using light wrought-iron spokes, from Joseph Glynn's treatise on water power.

tion were already starting to exceed the 20 feet or so that had been the usual limit to the diameter of wheels. John Smeaton built a few, including one as large as 47 feet. However, as Smeaton himself pointed out, wheels of this size, often still made of wood, were so heavy, unwieldy and slow that the benefits of size were limited. The lighter construction of the iron suspension wheel enabled the larger wheels to come into their own during the Industrial Revolution as part of the effort to extract extra power from the water.

Mining especially took to large wheels. They were a means of gaining additional power from what was often a limited supply of water, and wheels of 30 to 50 feet in diameter became common in mining districts. They were also a means of rationalization. It was quite common at Cornish mines to replace several small wheels with one large one. The mines of Cornwall continued to be the home of large wheels built in the nineteenth century. The largest, built at Boswedden mine, St Just, in 1837, was 65 feet in diameter, and several more in the county were

A drawing of a tin mine at Botallack, Cornwall, showing one of the large waterwheels used for draining and working ore. (Museum of English Rural Life)

greater than 50 feet. Bampfylde mine on the River Mole, Somerset, had a wheel that was 50 feet in diameter, and another of 35 feet. There were also some very large wheels at the Coniston copper mines in Cumbria. Even in Ireland, which was not a major mining area, there was a wheel of 40 feet, built in about 1750 for pumping a pit at Drumglass. Wheels used in mining were not only big, they were also numerous. There were thirteen wheels at the Coniston mines in the mid-nineteenth century. At Mary Tavy, Devon, the seventeen overshot waterwheels that were built for the Wheal Betsy and Wheal Friendship lead and copper mines represented the largest concentration of waterwheels in Britain. Eight of them pumped water, four wound ore to the surface and others worked crushing and stamping machines. The largest of the wheels was 51 feet in diameter, producing 87hp, and the smallest was 32 feet in diameter. The total fall of water to operate these wheels was 526 feet. Spread over a wider area, the mines of Devon Great Consols, on the Devon–Cornwall border, built up as many as thirty waterwheels from the opening of the mines in 1840.[180]

The biggest wheels were built in the mid-nineteenth century. The best known is the Lady Isabella wheel of 72 feet 6 inches diameter at Laxey, Isle of Man. It was built in 1854 to pump water from zinc and lead mines up to 2,000 feet deep, and survives still in preservation. Wheels of great diameter were not confined to the mines. The largest wheel built in the British Isles was one of 80 feet at a paper mill near Dublin, and there was another very large wheel at Darkley flax mill, Armagh, built in about 1850 and 70 feet in diameter. These large wheels were capable of producing up to 200hp. In England, Egerton Mill near Bolton was built in 1828–30 with a wheel of 62 feet diameter. It was built by William Fairbairn after its original designer, Johann Georg Bodmer, had returned home to Switzerland in poor health. This wheel could produce 110hp, and became a local attraction to such an extent that Ashworths, the mill

The great wheel at Laxey, Isle of Man, quickly became a tourist attraction, as this postcard of the early twentieth century shows. (Mills Archive Trust)

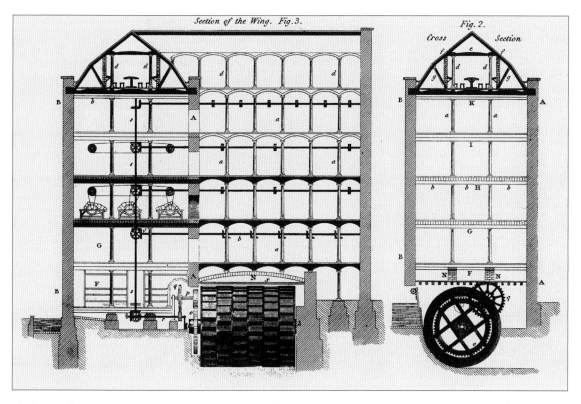

The drawing of the North Mill at Belper reproduced in Rees's Cyclopaedia *shows the wide, segmented wheel and the transmission system by which drive from the pitwheel is taken to a vertical shaft, from which overhead line-shafting runs through each floor. (Museum of English Rural Life)*

owners, kept a visitors' book alongside it. The Egerton wheel cost £4,800 to build and the wheel at Darkley £1,500; such high costs effectively limited the numbers of very large wheels.[181]

Wheels could also be made wider, and this was one of the principal contributions of William Fairbairn to the development of waterwheel technology. With all-iron construction, a wheel could even be made wider than its diameter.[182] The new wheel at Quarry Bank was 21 feet wide, and wheels with a width of 20 feet or more became quite common in the textile industry. Such wheels were, however, dwarfed by some of those at Strutt's mills at Belper. The breast-shot wheel in the West Mill was 48 feet wide by 12 feet 4 inches in diameter, and sufficiently remarkable in its dimensions to be described in Abraham Rees in his *Cyclopaedia* of 1819. To reduce the effect of impact on such a wide wheel, it was divided into six sections, with the floats at staggered intervals across the width. This principle of staggered floats to ensure smoothness to the drive was repeated in a similar wheel installed in the North Mill in 1802. This wheel, illustrated by Rees, was just 18 feet wide. The axle had to be constructed in sections, 'hooped together like a cask'. As if one enormous wheel was not enough, the West Mill had a second wheel of 18 feet diameter and 40 feet breadth as a standby for flood water. These wheels were made of wood, but after a few years they were replaced by iron wheels, still on the large scale. In place of the wooden flood wheel in the West Mill, a pair of iron wheels 21 feet 6 inches in diameter, with a combined width of 30 feet, were installed.[183]

Wide wheels had more general application than those of very large diameter. Whereas 8 to 10 feet had hitherto been the most common width, wheels of about 15 feet were used in a wide variety of mills in the nineteenth century.

Wide breast-shot wheel, Lower Slaughter, Gloucestershire. (Museum of English Rural Life)

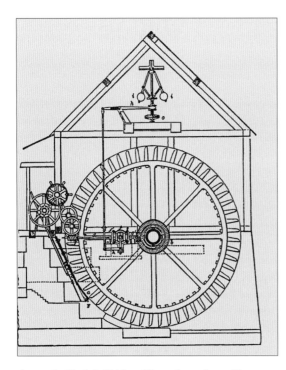

A waterwheel built for Waltham Abbey ordnance factory. The governor at the top regulated the speed, and this in turn operated a drive from the hub of the wheel to raise or lower the sliding hatch, thus ensuring as smooth a rotation of the wheel as possible.

Sometimes multiple wheels were employed to make use of a high fall. Joseph Glynn mentions two wheels, each 40 feet diameter, being installed at a mine near Aberdare, south Wales. They were placed one above the other, and revolved in opposite directions, the water flowing from one to the other in a figure of eight. A similar, and similarly unusual, example was built at Lumbutts Mill, Langfield, near Todmorden, in about 1833. This cotton mill used three overshot wheels, each 30 feet in diameter, arranged vertically one above the other, and housed in a tower wheel-house. There were three large reservoirs as well to feed the wheels. The method of operation is not clear now that the wheels and gears have gone, but it seems that water was re-used from the top wheel down to the bottom one. The maximum power from these three wheels was 50hp.[184]

There was an increase in the speed at which wheels were worked, but a smooth motion was more important, especially for the textile industry, where a sudden change in motion could break the threads. The speed of undershot wheels was determined by the current, but the breast-shot or overshot wheel

Lumbutts Mill was built high in the Pennines above Todmorden, Yorkshire. To capture the water supply this tower was built in about 1830 to house three waterwheels, one above the other. The wheels were fed from water collected in reservoirs higher up, but the precise workings of the system are unclear.

could be regulated more easily, and control was improved by the addition of the governor, introduced in the mid-eighteenth century. There were a number of ways of operating a governor. One involved the use of a pair of bellows tripped by a cam from the shaft of the waterwheel. If the wheel turned too quickly the bellows were inflated until they could take no more air and the wheel was slowed down. The centrifugal flyball governor was the type that became standard. Two or three balls rotated around a vertical shaft; as the speed of the engine increased, the balls were thrown upwards and outwards until they reached the horizontal, which forced the engine to slow down again. Linkages to the sluice gate could alter the inflow of water to the wheel. In conjunction with the sliding hatch, considerable refinement became possible in the control of the wheel's speed.

The flyball type of governor was first used to regulate the speed at which windmills worked. Thomas Hewes was credited with introducing the governor to the watermill, although there may have been precedents in the 1770s–80s. James Watt later applied it to steam engines, with which it also became standard equipment. Smeaton had found that a speed of 3½ feet per second at the circumference was the optimum at which a wheel should turn. A higher speed reduced efficiency. Nineteenth-century practice favoured speeds of 6 feet per second, thus increasing the revolutions of a wheel of 30 feet diameter from 2¼ to 4rpm. The greater use of iron in the construction of wheels was one of the causes of this. One of the consequences was that a waterwheel turning more quickly reduced the required ratio on the gearing to drive the machinery; alternatively, the machinery could be run faster, with the stones in a flour-mill gearing typically running at 120rpm.[185]

The new industrial era prompted a greater interest in measuring the power of engines and machines. Whereas the power of pre-industrial waterwheels has to be estimated from a small amount of information, millwrights and engineers of the late eighteenth and nineteenth century had a much clearer idea of the power they wanted.

The concept of comparing the power of a mechanical device with what a horse could do was an old one. Thomas Savery had applied it to his early steam engines and James Watt had introduced the term 'horsepower' to describe the power output of his rotative steam engines. He defined horsepower as the power required to lift 33,000lb 1 foot per minute. It was a nominal measure that underestimated the advantage that steam had over horses, to make sure that Watt's customers did not complain. However nominal, the new unit was soon being used to measure the power of waterwheels as well. Calculating the power produced by a wheel was based on the fall of the water and its flow. The fall was the distance the water descended in doing the work. On an undershot wheel this might be very slight, whereas for an overshot wheel it was only a foot or two less than the diameter of the wheel. The flow was the cubic measure of the amount of water passing through the millrace. For a basic calculation of horsepower, the formula was 'Flow × fall × 62⅓ divided by 33,000', where 62⅓ was the weight of a cubic foot of water, and 33,000 was James Watt's definition for the weight lifted.

Assumptions had to be made about the efficiency of wheels, but for the most efficient breast-shot and overshot wheels, a fall of 1 foot per second could produce 1hp from the wheel from 12 cubic feet of water. It might seem a rough and ready calculation, but, applied to many of the new wheels for industry, it proved a reasonably effective measure. The greater the flow of water, of course, the greater the power. Large rivers, therefore, would produce more power: the full flow of the River Bann in northern Ireland resulted in 3.3hp per foot of fall according to Fairbairn's calculations.[186]

Knowledge of the power produced by the waterwheel was important in the construction of the large factories. In broad terms, 1hp was needed for every 100 spindles in a cotton mill. A corn mill was reckoned at the end of the eighteenth century to need 8hp to turn a pair of stones. Most wheels produced 20–30hp, hence the spinning mill was most likely to have about 2,000 spindles, the corn mill three or four pairs of stones. To build a bigger mill required either an extra wheel, or a bigger one, such as the wide wheel at Quarry Bank.

During the nineteenth century the watt, which had been adopted for the measurement of electrical power, was extended to other machines. Its advantage was that it was a recognized standard unit, whereas horsepower had a number of variants from country to country. The power of water turbines was regularly expressed in terms of the kilowatt or megawatt, especially once they were used to generate electricity. This usage was extended to the older waterwheel to some extent. In Britain the kilowatt is equivalent to 746hp.

The siting of waterwheels could change. The tradition was always to place the wheel parallel with the stream, and most continued to be positioned in

Sir William Fairbairn's diagram of the construction of an underground tailrace, which enabled the water to drop cleanly away from the wheel. (Museum of English Rural Life)

this way. This was determined largely by the construction of the water channels to and from the wheel, which generally ran alongside the mainstream. The tailrace running parallel discharged back into the river after a short distance. The wheel at Quarry Bank was one of a number built more or less at right-angles to the river. That provided more convenient arrangements of drive to the machines. To position the wheel thus required the tailrace to be dropped below river level at that point and taken in a tunnel to rejoin it much further downstream. William Fairbairn built some wheels arranged in this way: it enabled him to install a larger wheel with greater fall, the gain in power being evidently sufficient to offset the greater cost of building the tailrace.[187]

Fairbairn also favoured a concentration of power generation on the best site available, in contrast to the prevailing practice of placing the wheels as close as possible to where the work was required. In a small corn mill this was of little consequence, but at a large textile mill it could result in three or four independent waterwheels. Fairbairn's preference, which he put into practice at Catrine and Deanston, was to place the wheels together in their own wheelhouse, thus maximizing the power from a uniform fall. This more than made up for any loss of power in transmission across the site, for gearing was simplified too.[188]

THE INTRODUCTION OF IRON

Before the eighteenth century, mill wheels and gearing were made of water-resistant hardwoods. Oak was the most common, being usually the cheapest and most abundant material. Elm and hornbeam were other frequently used woods, while beech, plane and hickory might find a place occasionally. These were durable materials: it was reckoned that a good oak wheel might last for up to eighty years, and in ideal conditions with a moderate rate of use that might be the case. There were, however, many aspects to the use of wood that could shorten a wheel's life, so that twenty or thirty years was a far more realistic life expectancy. The alternate

The wooden breast-shot wheel in Longbridge corn mill, Hampshire.

wetting and drying as the wheel turned could cause serious problems. Planks expanded then shrank again, causing joints and pegs to work loose and the wheel to get out of balance. Some parts might rot. Not all timber was of a consistent strength, and stress and strain might cause splitting. The oak axles, which had to bear the load and transmit power, were especially prey to such problems and many did split.

The rotting and splitting meant that wooden wheels required regular attention to replace worn parts and, eventually, the complete wheel. The average life of a wooden wheel was probably nearer to twenty years and many needed replacing after about ten. As Quarry Bank Mill at Styal expanded, the wheel that was in use in 1784 was replaced in 1792, and this in turn was replaced in 1807 by an iron wheel. A second wheel of 1800 was replaced in 1820. At Tutbury, the wheel installed in 1781 lasted until 1800–10 when an iron wheel was put in. Of course, a short working life was not necessarily just because the woods had ceased to be as durable, but was also a result of the continual search for extra power and efficiency in the mills of the Industrial Revolution, with mill owners being prepared to invest in new, more powerful wheels. The wheels in textile mills were likely to be subject to more continuous use than those in corn mills, and thus needed more regular maintenance.

There was a limit to the size and power that could be achieved with a wooden wheel. A diameter of 40

Developing the Technology of Water Power

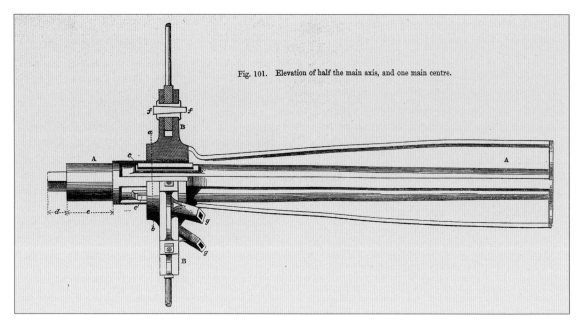

The construction of a cast-iron axle for a waterwheel. (Museum of English Rural Life)

feet was about as big as they could get. The members of the wheel had to be of solid timber in order to give it strength. Wooden floats and buckets had to be thick, which restricted the number of vanes and inhibited the entry of water on to them. For a mill owner of the eighteenth and early nineteenth century who was looking to increase the power and efficiency of his waterwheels, iron construction was an increasingly attractive option.

Iron had been used for some parts of waterwheels and mill gears for centuries, but because it was expensive and limited in supply it was used sparingly. The introduction by Abraham Darby of coke as a fuel for the smelting of iron ore, in 1709, one of the 'classic' stages in the Industrial Revolution, led to an increase in the supply of iron and a fall in its price. The greater supply and lower price of iron made it a more practical proposition for millwrights.

It took some time for these developments to affect the use of iron in mill work, which was still limited in the mid-eighteenth century. There was some iron used in the London Bridge Water Works, and there was a mill in Arbroath which used iron teeth on gearing, but such examples were still isolated.[189] John Smeaton has generally been credited with the breakthrough of introducing cast iron for the shafts and axles of waterwheels. He found it almost impossible to make wheels to his standards using wood, and had to build some of his early wheels in sections, tying them together with iron plates. He did that on the first wheel for which he used iron, at Wakefield Mill in 1754.[190] He tried cast iron for an axle first in a windmill in 1755. In 1769 he recommended cast iron for a new shaft to replace one of oak that was failing on a waterwheel at Carron Ironworks, for which he acted as consultant, and this was acted upon in 1770–1.

Cast iron can be brittle, and some of Smeaton's early wheelshafts failed as a result. Enough were successful, however, for him to continue to use iron, and to extend it to other parts of the wheel. He used iron for the rings of the wheel for the furnace-blowing engine at Carron in 1770. He next introduced wrought iron for waterwheel buckets, in 1780 at a mill in Carshalton, Surrey, which also increased their capacity.[191] He also used cast iron for gears, first at Kilnhurst Forge in Yorkshire in 1765, at Carron for the spur wheel in 1768, and at Ravensbourne, Deptford, in 1778. Smeaton's example was followed by other leading engineers,

This iron wheel was used at the corn mill, Cromford, Derbyshire.

including John Rennie, who adopted iron for gearing in the 1780s. Gradually, the possibility of achieving lower friction, greater efficiency and higher productivity led to the greater use of iron.

The next logical step was to build the waterwheel entirely of iron, but it was not until about 1800 that this was done, and it is not certain which was the first. Among the contenders are the wheel built by Peter Ewart at Quarry Bank, and one by Hewes at Belper, dating from the first years of the nineteenth century. A cast-iron wheel was installed by Watkin George at the Cyfartha Ironworks, South Wales in

An iron pitch-back wheel at Castlebridge Mill, Wexford, in 1938. (Museum of English Rural Life)

Developing the Technology of Water Power

A wheel of hybrid wood and iron construction. This is a survivor from the mine workings of Swaledale at Pillmire Bridge, near Reeth in Yorkshire.

1800. This large wheel, 50 feet in diameter, was used for blowing blast furnaces.[192]

It took some time for iron to supplant wood and many waterwheels of the late eighteenth century were still being made either wholly or mainly of wood. The wheel for Belper West Mill of 1795–7, which cost £629 19s 4¾d, was made almost entirely of oak.[193] There was a good deal of scepticism about iron, especially about the liability of cast iron to fracture. The first generation of castings from the Industrial Revolution were not always strong enough, and breakages were common. Also, although it had a longer life, iron cost more than wood, and there could be problems in obtaining iron parts and maintaining them. Local foundries were not always able to cope with larger parts, which meant that replacements had to be bought from more distant makers. This would lead to higher transport costs and, possibly, time lost with the mill out of commission waiting for the delivery. Many millers, especially in the small country mills, simply preferred wood, especially for the mill gear wheels. Local traditions of millwrighting, experienced in handling wood, helped ensure that wooden mill

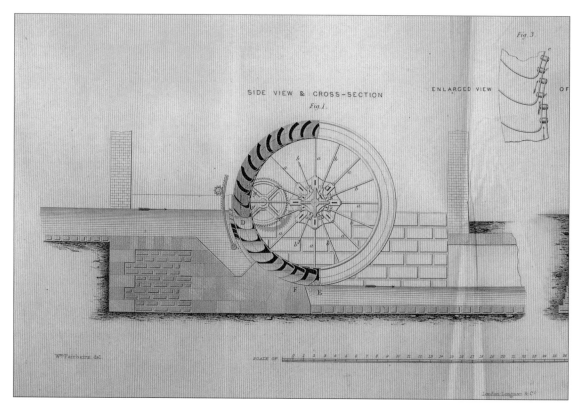

Fairbairn's iron wheels built for the cotton mills at Catrine perfectly illustrate the suspension-wheel construction, and the drive from rim gear to a larger transmission gear, which in turn engages with gear for the machine shafts. (Museum of English Rural Life)

Developing the Technology of Water Power

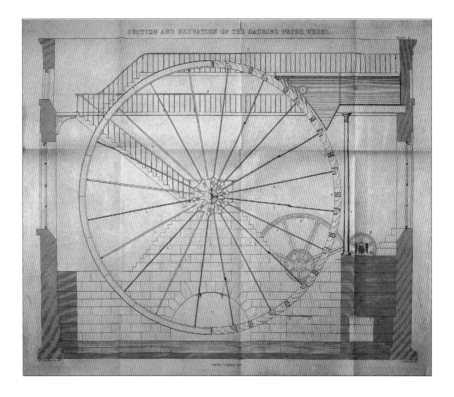

Sir William Fairbairn's drawing of one of his improved low-breast waterwheels built in the late 1820s, showing the construction of the masonry breast and the iron gear wheel that takes the drive from the rim gear. (Museum of English Rural Life)

gearing continued in use throughout the nineteenth century, and much still survives. Wallowers, spur wheels and other drive gears were very likely to be made of wood and even when they were made of iron, some preferred to have oak cogs. Iron was used more commonly for the pitwheel, which would be cast in one piece, but split into two halves for ease of transport and installation.[194]

The waterwheel was more likely to be made of iron, but all-iron wheels were not universally popular, and composite construction was common well into the nineteenth century. The expense of iron and the belief that wood was better able to take some of the stresses meant that millwrights would commonly use wood for the spokes alongside iron for the rims. Sometimes, a wheel had wooden buckets and a hub and spokes of iron: the wheel of about 1850 at Higher Mill, Helmshore, was of this construction. Many of the very large wheels of the mid-nineteenth century, including the Lady Isabella wheel on the Isle of Man, were of hybrid construction.

One of the main weaknesses of the traditional waterwheel was that the main shaft carried the whole of the load of the waterwheel and the pitwheel, and at the same time acted as power transmission. The solution to this was to take the power off in another way, make the axle only slightly longer than the width of the wheel, and support it in axle-boxes at each end, with bearings to reduce friction. This was the principle behind the suspension wheel. Shaft and arms now only had to carry the weight of the waterwheel. In consequence, wheels could be of much lighter construction. Wooden wheels could be built in this way, and a few were. However, it was the introduction of iron, with its greater strength and lower friction, that enabled such wheels to come into their own.[195]

The means of power take-off with the suspension wheel was the rim gear. Cogs set in the periphery of the waterwheel itself engaged with a small gear, from which power was transmitted to the machinery. This again was not a new concept: it had been known since the sixteenth century. Teeth in the rim of the waterwheel could engage with a lantern wheel. However, the peg-style teeth in general use at the time were always liable to become dislodged; being attached to the waterwheel and getting alternately wet and dry could lead to maintenance problems. Millwrights,

Developing the Technology of Water Power

New wheels for old at Cobham Mill, Surrey, 1953. (Museum of English Rural Life)

therefore, preferred the separate pitwheel as the main drive from a wooden waterwheel. Iron gear teeth were not susceptible to these difficulties, and the greater strength in general of the material made it more suited to rim gearing.

The iron suspension wheel with rim gearing was introduced in about 1800. Thomas Hewes, a millwright from Manchester, is generally reckoned to be the first to have built one. He was responsible for the iron wheels with rim gearing installed at Belper for Jedediah Strutt between 1804 and 1810, which enabled simplification of the drive transmission to the textile machinery.[196] Suspension wheels were able to make greater use of the tensile strength of iron. The first iron wheels followed the pattern set by wooden wheels, with solid radial spokes. Thomas Hewes introduced light wrought-iron tension rods that braced the rim to the hub in the same fashion as on a modern bicycle wheel. Hewes's first wheels of this design were probably the two of 15-foot width that he built for Belper in 1811. Other engineers followed Hewes's lead in the use of iron suspension wheels and William Fairbairn in particular refined the design for his industrial wheels.

Together, the suspension wheel and rim gearing enabled larger, more powerful wheels to be constructed, and this was a key element in the increased water power available for nineteenth-century industry.

The standard of waterwheel design and construction was undoubtedly improving by the early nineteenth century, under the influence of the new generation of engineer-millwrights following in the footsteps of John Smeaton, and the introduction of new techniques. In the mid-eighteenth century many, perhaps most, wheels were of indifferent quality. The engineer John Rennie conducted a tour of corn mills around the country in the 1780s, making notes on the many that had potential for improvement. He was not impressed by one at Newbury: it was 'so very ill-constructed' that it caused the owners to 'suffer at least twenty-five per cent of the water to run to waste'. A millwright in Salisbury in 1810 commented that 'in most instances all wheels are made of the same diameter and width without the least attention being paid to the quantity of water or height of fall'.[197] When William Fairbairn was called upon to replace the wheels at the Catrine Cotton Works in 1824 he reported that the existing wheels were 'ill-constructed, deficient in power and constantly breaking down or getting out of repair'. They had been installed forty to fifty years previously by Mr Lowe of Nottingham, one of the leading millwrights of his day, and had thus lasted relatively well. Fairbairn accorded Mr Lowe faint praise: 'His work, heavy and clumsy though it was, had in a certain way answered the purpose, and as cotton mills were in their infancy, he was the only person qualified from experience to undertake the construction of the gearing.'[198]

Fairbairn was very proud of his own work. The wheels he put in at Catrine were, he claimed in the 1860s, 'even at the present day among the best and most effective structures of the kind in existence'.[199] His pride in his work was certainly justifiable: the wheels worked well for 120 years. Several of the wheels built at this time were similarly long-lived, a measure of their quality. A life of fifty years or more was common. Even then the wheels were not necessarily worn out, but were being replaced by steam, electricity or, as in the case of the Catrine wheels, a water turbine. Among the very long-lived wheels, the one at Deanston cotton mills worked from 1831 to 1949, and that at Arkwright's mill in Bakewell was installed in 1827 and worked for 128 years, until 1955.

It was not a straightforward matter to get sufficient fall of water to drive an overshot wheel, which needed a minimum fall of 10 feet. The breast-shot wheel was a possibility where sufficient fall was not available, as it could be worked with a fall of between 4 and 10 feet. In the lowland districts, achieving such fall was difficult: the River Stour in Suffolk had an average fall of 6 feet 3 inches per mill along its course in the 1750s. For any mill in these circumstances to improve its fall, therefore, required some extensive, and expensive, civil engineering, and it was likely to encounter some of the difficulties from other mills along the river. Some millers did decide it was worth the effort.

FROM MILLWRIGHT TO ENGINEER

The technological developments of mill wheels and gearing changed the nature of the millwright's work. Millwrights were already starting to call themselves 'engineers' by the 1750s, although some, such as J. T. Desaguliers, felt that this was not justified since they lacked even a basic scientific knowledge. By the

The corn mill at Nuneaton with two pairs of stones driven from the one waterwheel, illustrated by Henry Beighton in 1723. (Museum of English Rural Life)

Developing the Technology of Water Power

nineteenth century, however, the true millwright-engineer was emerging, of whom Sir William Fairbairn was a prime example.

Fairbairn himself argued that the millwright was a man of substance, educated and well versed in 'mill machinery, pumps and cranes and could turn his hand to the bench or forge with equal adroitness and facility'. Most millwrights probably did not fit this description, but leaders of the profession, such as Fairbairn himself, certainly did. Such men were leading a new type of industrial millwrighting and engineering business, and the most important of these in the early nineteenth century was that created by Thomas Hewes, the builder of the first iron suspension wheels. He moved to Manchester from Belfast in 1792, and designed many textile mills of the late eighteenth and early nineteenth centuries, supplying the wheels and gearing. As well as pioneering the suspension wheel at Belper, he also built the iron wheel for Styal in 1807, and its successor, the 100hp wheel of the 1820s. By the 1820s, Hewes's was the largest firm of millwright-engineers in Britain, with about 150 employees.

William Fairbairn came to work for Hewes in 1817, but left after only a few months, returning home to Scotland, where he went into partnership with James Lillie as millwrights and engineers. He improved upon Hewes's designs for suspension wheels, enabling them to run more smoothly. In the 1820s he installed large wheels for the Catrine Cotton Works in Ayrshire and the Deanston works in Perthshire. They were high breast-shot wheels of great longevity, lasting until turbines replaced them in 1947 and 1949 respectively. Between 1820 and 1851, Fairbairn was reputed to have supplied about 300 waterwheels, some to overseas customers. He also wrote one of the nineteenth century's most influential treatises on their construction.[200]

POWER TRANSMISSION

The first successful attempts to drive more than one pair of millstones from a single waterwheel were made probably in the late seventeenth century. Henry Beighton, one of the foremost land surveyors and engineers, based in Warwickshire, published an engraving in 1723 of a watermill for grinding corn at Nuneaton, which clearly shows two pairs of stones being driven from the one overshot wheel. Power was transmitted from the waterwheel to the vertical cogwheel. Beighton then shows two separate drives from the cogwheel. One is the conventional lantern wheel. The second drive is to a vertical gear at the 3 o'clock point on the cogwheel, which is depicted as having a slightly larger diameter than the trundle. A horizontal shaft extends forwards from which another vertical drive transmits power through a lantern wheel to the stones above. Beighton shows only one set of stones driven from the horizontal shaft, but it could be extended further, and any number of drives taken off it. Other corn mills built

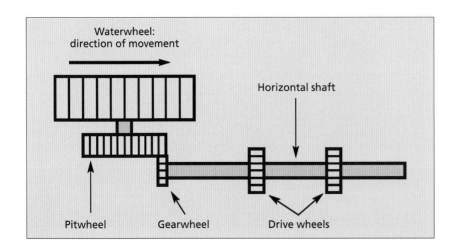

Plan of the layout of drive for a horizontal shaft, as depicted by Beighton.

Developing the Technology of Water Power

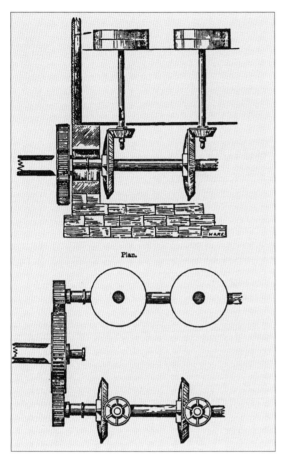

The layout of a typical layshaft drive running in line with the axle of the waterwheel, from Joseph Glynn's treatise on water power.

Layshaft drive in Hockley Mill, near Winchester, Hampshire. This was a farm mill built to drive a threshing machine and corn-grinding stones from the one horizontal shaft.

to this layout are known with three pairs of stones, one from the upright lantern gear and two from the horizontal shaft (*see* opposite).

Although Beighton's engraving dates from 1723, it is likely that this type of drive was devised earlier, during the late seventeenth century. Ruswarp Country Mill in north Yorkshire seems to have had a similar drive by about 1682, and there are indications that other mills might have had drive to two or three pairs of stones before 1700.[201]

Drive from a horizontal shaft was the breakthrough, enabling grain millers to operate several pairs of stones from one waterwheel during the first half of the eighteenth century. Known generally as 'layshaft' or 'lying shaft drive', it also formed the basis for many other industrial users of water power.

Later in the eighteenth century it became more usual for the horizontal shaft to extend along the same line as the waterwheel's axle instead of at right-angles to it. The vertical lantern drive was then usually dispensed with. At about the same time as drive-through horizontal shafting was being introduced, the intermediate gear from the waterwheel, the cogwheel, became known as the pitwheel, because it was usually sunk below the ground floor of the mill.

Layshaft drive was a boon to the grain millers, for it offered a solution to an issue that was becoming increasingly important: the need to grind different types of grain. There was a growing demand for white wheat flour, which was best produced with fine, hard millstones. British millstones were mainly of coarse sandstone or millstone grit, most often sourced as 'greystones' from the Derbyshire Peak

Developing the Technology of Water Power

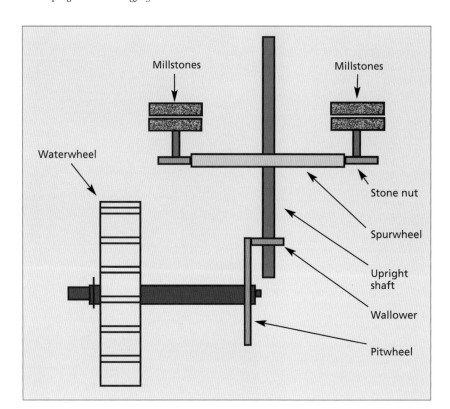

The basic arrangement of drive in a corn mill using an upright shaft and spur wheel.

District. For best-quality wheat flour, stones were imported: from Germany came 'blue' (or 'cullin') stones of volcanic stone, and from France came the 'French burr' stones, a type of sandstone. French millstones had been imported since as far back as the fifteenth century. Blue stones had become popular from the late seventeenth century onwards, especially in the north-east of England, but French burr stones were ultimately the most popular for grinding wheat flour. Most stones had been cut from

Gearing in Sarehole Mill, near Birmingham, a mixture of iron and wood construction.

the rock in one piece, but French burr stones were usually made up from several smaller pieces, bonded by cement and secured by an iron band.[202]

As grain millers from the late seventeenth century onwards were trying to meet increasing demand for different products – fine white flour, wholemeal and oat flours, and meal ground for livestock feed – they were keen to operate more than one pair of stones driven by the one waterwheel. The new layshaft drive allowed them to do this, and subsequently to add further machines, such as the bolting machine, which cleaned the white flour after grinding.

One alternative to layshaft drive, developed at about the same time, also enabled more than one pair of millstones to be worked off the one waterwheel. This used a vertical main drive shaft (always known as an 'upright shaft') and a new horizontal gear, the 'spurwheel' (or, often, the 'great spurwheel', because of its large diameter, which was commonly 6 to 12 feet). The spurwheel worked in a fashion similar to the large horse gears of the seventeenth and eighteenth centuries, which involved the horse walking below a gear from which drive was taken down to the machine. Indeed, it is quite possible that this was the inspiration for the spurwheel. In a grain mill, spurwheel drive involved the pitwheel engaging with a small gear, the 'wallower', which replaced the lantern wheel. Above that was the spurwheel, and smaller pinions could engage from its rim to drive separate pairs of millstones. The gear ratios on the different pinions could be set to enable the miller to operate the different stones to suit different flours. The spurwheel, usually of wood and often of clasp-arm construction, needed to be solid and heavy.

If the waterwheel was powerful enough and the spurwheel large enough, up to eight pairs of stones might be driven in this way; more stones would require another waterwheel. Most corn mills in fact had no more than three pairs, with just a small proportion having four or five pairs. In the most usual arrangement, known as 'underdrive', the spurwheel was below the stone floor ceiling, with drive shafts running up to the stones on the floor above. Less often, the spurwheel was on the stone floor, to

The pitwheel and drive in Marcham Mill, near Banbury, 1953. (Museum of English Rural Life)

give 'overdrive' transmission; this was probably introduced somewhat later.

Using drive from the main upright shaft led to a change in the layout of watermills: the stones, which had often been to the side of the building, were now grouped together in the centre of the floor. The upright shaft also suited many flour mills better than layshaft drive because it could be extended through to upper floors and used to drive additional equipment. Initially, the main use for the power was to operate the sack hoist, driven from a gear wheel on the upright shaft called the 'crown wheel'. It is first mentioned in the eighteenth century – by John Smeaton, among others, who refers to it as powered, not manual. At about the same time mills were starting to use bolting machines to sift out the bran after milling, and these, too, were driven from the waterwheel. In the nineteenth century, other pieces of machinery, such as corn screens and grain elevators, followed suit. A few mills had silk reels, which enabled flour to be finely graded – Tocketts Mill, near Guisborough, Yorkshire, still has one – although they were more usually found in the larger steam mills. All of these machines could be driven by water power through the drive shaft and gearing.

Spurwheel drive was first used in windmills, the gear fitting well into the confined space. Chesterton windmill, Warwickshire, built in 1632, is the first recorded example. It is uncertain when this arrange-

Developing the Technology of Water Power

A bolting machine for sifting and grading flour, of the type installed in mills from the nineteenth century onwards, with drive take from the waterwheel. This was in Ewell Mill, Surrey. (Museum of English Rural Life)

ment was first applied to watermills, but it was becoming established by the early eighteenth century. Spurwheel drive became the most common form of gear transmission for corn mills, but it took some time before this was achieved. It was not until the late eighteenth century that spurwheels started to appear in the watermills of north Yorkshire, for example, demonstrating that it was increasing demand for flour and meal from this time onwards that was prompting the rebuilding and re-equipping of many corn mills.[203]

Henry Beighton's drawing of the Nuneaton mill shows the continued use of gears of tooth and peg construction, which were then still the most common form. Bevelled gear wheels, with notched and angled cogs to engage with one another, were a development of the mid-eighteenth century, and began to become the standard during the Industrial Revolution.

Spurwheel drive was applied to some other industries that involved similar grinding or crushing operations. In other industries, drive via horizontal shafting was more common. This might involve drive to a layshaft running under the building. Alternatively, drive from the pitwheel or rim gear was transmitted through intermediate gears or belts to line-shafts running the length of the building. The use of light wrought iron for the line-shafts, introduced in 1817 by William Fairbairn, encouraged this means of working. Such continuous shafting from the waterwheel enabled as many drives to machines to be taken as were needed, with gearing stepped up or down to suit. At Arkwright's cotton mill at Cromford, power was taken from the waterwheel to the machines by a system of wooden drums and leather belts. As the new textile industries grew, power transmission systems were refined. William Fairbairn played a major role in this, working with James Lillie to formalize standards for power transmission during the 1820s.[204]

WATER SUPPLY

The Industrial Revolution increased demand for water power and that in turn increased the pressure on water supply. The new factories consumed power and water in amounts far in excess of the corn mills and other pre-industrial users. Arkwright's mill at Shudehill consumed 6,000 gallons a minute, falling

A bevelled gear wheel made of wood in Eling tide mill.

30 feet to drive 4,000 cotton spindles. The large wheels that Fairbairn built at Catrine took 70 tons of water a minute from the River Ayr. The 62-foot wheel at Egerton Mill in Lancashire needed 4,320,000 gallons of water a day to keep it at full power. This proved to be unattainable, as competition for water, and the mill owners' failure to secure undisputed rights to that water, starved it in dry seasons. The summer of 1842 was a particularly hot one, when many mills had problems. At Egerton the flow of water was so inadequate that the wheel could work only five whole days and forty-nine part-days in the whole of that year without supplementary steam power. That same hot summer closed the Ashworths' other mill at New Eagley for five weeks. Here the water supply had been reduced after the building of a new Belmont reservoir, which drew much of its water from the gathering grounds for the Eagley Brook on which the mill was sited. Given such potential problems, when the water was flowing well, the mills were likely to work extra hours to make up lost time. The Ashworths, being among the better employers, were generous enough to allow holidays for their workers when the water was low.[205]

It was not the new industries alone that put pressure on water supply. Demand was increasing from smaller users as well. Corn mills wanted more power to drive extra pairs of stones, and the various small water-powered workshops in metal, wood and other trades crowded on to streams. By the beginning of the nineteenth century examples of crowded streams abound throughout the British Isles. There were forty-six watermills within a 4-mile radius of Perth by 1800. The number of mills in Galloway doubled between 1750 and 1850 and all the available sites in the Birmingham area had been taken by the 1760s. As the cotton industry expanded in Lancashire, streams became crowded with mills. In 1835 Edward Baines claimed that there were more than 300 mills on the River Irwell and its branches along the stretch between Bacup and Bolton. In this stretch of about 4 miles Baines reckoned there was a total fall of 900 feet, of which 800 feet were already occupied by the mills engaged in the spinning, scouring, bleaching, printing and dyeing of cotton. The Spodden, a tributary of the River Roch, to the west of Rochdale, supported eighteen mills in a course of 9 miles, and this density was replicated along almost every stream in the area. Professor Unwin wrote of there being a race for water power at Stockport in the 1790s. One effect was to force manufacturers to cast their net wider in their search for a site for a mill. As a result, the new textile industries were more widely dispersed than they might otherwise have been, with cotton mills being built at such places as Flintshire or Furness, for example.

Ultimately, inadequacy of water supply was to be the downfall of water power in a number of instances, but as long as steam power was not providing effective competition, mill owners took a number of steps to improve matters. In the first place, the rights to use water assumed greater importance as more mills crowded along rivers. It has been seen that uncertainty over rights might well have been one reason for Arkwright's choice of the Bonsall Brook over the Derwent as power source for his first mill at Cromford. It was certainly the reason for another Derbyshire mill of the 1780s, Cressbrook, which would draw power from the little Cress Brook rather than the River Wye, over which the Duke of Devonshire already owned the rights. Restrictions on water rights was also one limiting factor to the use of water power for the expanding silk industry at Macclesfield.[206]

Water rights were the property of the owners of the neighbouring land and river bed. They were rented to mill owners in the same way as a piece of land, except that it was not always as simple. Matters were complicated by the competing claims of other users of the river and its water, and by the fact that, in areas where watercourses were already a complexity of streams, mill leats and drainage channels, it was not always clear precisely which stream was being drawn upon. It all made it less easy for a mill owner to gain absolute rights to draw off all the water he wanted. With large numbers of mill owners, mining interests, boatmen and fishermen competing for rights, rents were likely to increase, and so was litigation. In the early nineteenth century it was reckoned that legal costs of disputes over water in Lancashire and Yorkshire could be as much as £10,000 a year. In 1846 R. H. Greg of the Styal cotton-spinners told a parliamentary committee that manufacturers must 'either pay a high rent for their water power, or they have a very great outlay in improving it, so that it is very expensive'.[207]

A second way of securing an adequate supply of water was to improve the mechanism to capture and store it. Sometimes it was a matter of enlarging a pond a little to build up some extra supply overnight when the mill was not working, but many went much further. Builders of medieval and early modern mills had produced some long and elaborate watercourses to serve them, but they were surpassed by many engineers of the Industrial Revolution, who built and rebuilt systems to supply enough water for their large overshot and breast-shot wheels. Weirs were made bigger: some were now large enough to be called dams; the reservoirs behind them were larger; and mill leats were longer. A prime example of a water management system to save and make the best use of limited supplies was built in the 1780s on the River Leen in Nottinghamshire. George Robinson had built cotton mills by the river, using some older water-power sites. The flow of the river at this point, however, was limited; in order to secure enough water, Robinson built a series of reservoirs, extending in total over more than 30 acres, to serve his six mills. In addition, he built a lengthy channel running parallel to the main river to serve two of the mills, carried on an aqueduct and high embankments in order to gain sufficient fall. The highest mill was supplied from a reservoir. From this mill some water was taken by an embankment and aqueduct to the next mill, Grange Mill, which had its pound to maintain a steady supply. The aqueduct was taken further to supply another mill, while the river directly supplied three further mills, each with its pond to maintain supply.[208]

Similar developments took place at almost every location where the new water-powered factories were built. The 'race for water power' at Stockport led to the construction of a series of dams, reservoirs and channels in tunnels. On a smaller scale, channels built up on trestles fed water into the reservoir at Millhouses, south of Wirksworth. At Belper, Strutt built his big horseshoe weir to impound the Derwent to the extent of 14 acres. After writing about the crowded nature of the River Irwell, Edward Baines went on to describe plans to build additional reservoirs, which would, he reckoned, provide enough water for 6,600hp. One of the eighteen reservoirs had been built by 1835, and another was under construction. Longfords Mill, Gloucestershire, had a reservoir of 15 acres held back by a dam 30 feet high and 450 feet long, but that was exceeded by the dam 108 feet high built at Entwistle for the Eagley mills. Sedgwick gunpowder mills near Kendal had a leat

Dane in Shaw Mill, outside Congleton, built in 1784, stands beside its large reservoir.

half a mile long built in 1852 to create a head of 18 feet and power potential of 150hp. The tailrace was in a tunnel 200 yards long. To serve the mines of the north Pennines, some large dams and reservoirs were built, such as the Burnhead Dam, 40 feet high and 500 feet long. A major reconstruction and mechanization of the lead mines of Grassington Moor in the late eighteenth century led to the construction of three large reservoirs, from which water was conveyed along a 6-mile network of channels, partly in a tunnel, to serve a dozen wheels.[209]

The nineteenth century saw some major water-engineering projects. Neighbouring mill owners would combine to promote schemes, and sometimes local authorities joined in, in order to secure water for urban use. Sir William Fairbairn was commissioned with J. F. Bateman in 1836 to re-work the flow on the River Bann in Ulster. The result was the enlargement of Lough Island Reavy, from a natural lake of 92½ acres into a reservoir of 253 acres. The smaller Corbet Lough was also enlarged and new feeder channels constructed. One of the largest water-engineering schemes of this period was at Greenock. Initiated in 1824, its purpose was to supply the drinking and sanitary needs of the town, but the consulting engineer, Robert Thom, realized there could be further benefits for the users of water power in the Glasgow area, through provision of enough storage for six months. Water for the city's consumption was diverted from Loch Katrine, and in compensation for users of water power a new reservoir was created by raising the level of Loch Vennaqar. This created a total fall of 512 feet, which could be tapped by a series of water-driven works.

Similar development was undertaken when Edinburgh sought to secure its water supplies: Glencorse dam and its reservoir, built between 1819 and 1822, provided for the local mill owners as well as for the city dwellers.[210]

Corn millers were just as desperate to secure additional water supplies. From the mid-eighteenth century onwards there was widespread rebuilding of mills and their water supply. Weirs were raised, additional dams added and new channels cut, some of considerable length taking water across watersheds from one stream to another. In the southern North York Moors, a channel 1½ miles long was cut to draw water from Loskey Beck to Hole Beck on which the mill at Lastingham stood, and there were a number of similar examples of new watercourses in this area.[211]

Although the number of works was reduced from the 1820s, as steam power began to make a greater impression on industry, there continued to be some construction of reservoirs and channels for waterwheels until the mid-nineteenth century. The demand for water from the growing number of textile mills led to some co-operative efforts in building large reservoirs to supply several mills downstream. Some landowners took the initiative in doing this: Lord Dartmouth was one. Sometimes the mill owners got together, as in the Luddenden Valley, a tributary of the Calder in Yorkshire. The owners formed the Cold Edge Dam Company in the early nineteenth century to build a reservoir on the moor of that name. By the 1830s the company had three reservoirs high on the moors supplying ten mills in the valley. Most were woollen and worsted mills, but

Developing the Technology of Water Power

there was also a corn mill and paper mill among them. Another group of mill owners formed the Holmes Reservoir Commission to build a large reservoir, the Bilberry reservoir, above Holmfirth, work authorized by Act of Parliament in 1837. The Kentmere Head dam, built for the millers of Kendal between 1845 and 1848, was among the last before a new phase of dam construction began to support hydro-electricity generation.[212]

The promoters of the Kentmere scheme had plans for further work, but were forced to cut back when they ran over budget. Costs of hydraulic engineering work were high, and ultimately prohibitive once steam power proved an effective competitor. In 1795–7 Strutt had spent £1,142 on the new cut for his West Mill at Belper. By the 1820s bills for large-scale reservoir works were in the tens of thousands, and there was no guarantee of security of water supply after that, as was discovered at Eagley. The Holmes Reservoir Commission sought to raise £40,000 to build Bilberry reservoir, but they lacked the capital they really needed. This shortfall may have contributed to the poor engineering of the dam, which burst in 1852, killing eighty people and destroying mills in the Holmfirth area.

WATER TURBINES

The water turbine is essentially a horizontal waterwheel powered by a jet of high-pressure water that turns it at high speed. It is smaller than the vertical wheel, more economical of water and low in maintenance. It can last for thirty to fifty years in service – some have worked for much longer – and it can also work with almost any head of water, great or small. Vertical wheels have problems both with very low head and very high, with an effective maximum of about 50 feet. Because the turbine runs at higher speed, the gearing needed for industrial uses is less complicated, and its speed is more easily regulated with a governor. Power is consequently much greater and produced more efficiently. All of these advantages made the turbine a very attractive proposition for many industries, helping to prolong the use of water power in many instances, while it became an important tool for the new industry of hydro-electric power generation.

The term 'turbine' was coined by Claude Burdin for fast-running wheels, but it soon came to mean specific types of wheel to generate power. Among their defining characteristics were their small size, rapid revolution and ability to work with any head of water, along with the fact that they could work when submerged in the stream.

The genesis of the water turbine is a long one: as early as the fifteenth century Francesco di Giorgio made a drawing of what was essentially a water turbine wheel. However, the main developmental work was undertaken during the eighteenth and early nineteenth centuries and involved the efforts of many hydraulic engineers, mostly in France. On continental Europe the horizontal waterwheel had retained a more important role than it did in Britain,

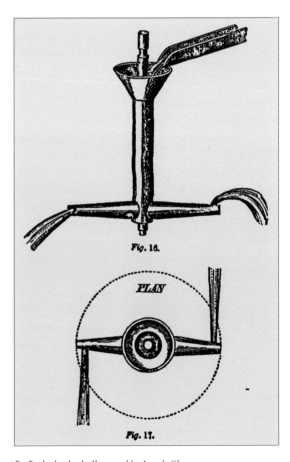

Dr Barker's wheel, illustrated by Joseph Glynn.

and engineers were interested in improving its efficiency and power output. At the same time as John Smeaton was conducting his experiments that led to British engineers adopting the more efficient overshot and breast-shot wheels whenever they could, French engineers were studying the motive forces on horizontal wheels. Their incentive was the greater abundance of water power compared with coal in France: if their country was to follow Britain's lead in industrialization, any increase in the efficiency of water power would be invaluable. British engineers may have been slow to take up the possibilities of the water turbine, but this is usually attributed to the rise to dominance of steam power in British industry during the mid-nineteenth century. However, the turbine did come to play a significant part in the development of water power in the British Isles.

The man usually credited with inventing the water turbine was Benoit Fourneyron, who designed a compact high-speed waterwheel with 80 per cent efficiency, which he installed at Pont sur l'Ognon, Puy de Dôme, in France between 1823 and 1827. He went on to construct about a hundred similar wheels. Among the features he introduced was an encasement of the wheel, which increased the water pressure.

There are two basic forms of water turbine: the impulse and the reaction. An impulse turbine works by directing water at speed from one or more jets on to the wheel. The exit passages had to be designed to clear the water smoothly, for, just as with a vertical wheel, tail-water falling back on to the wheel causes it to lose power. Behind this lay one of the great maxims of hydraulic engineering: water should impact on the wheel with no shock and leave it without velocity in order to achieve maximum efficiency. A reaction turbine works under the pressure of water. The water passages are completely filled, and the energy of the water at the inlet is transferred to the wheel as it passes through.[213]

The simplest reaction turbine works on the principle still used for a lawn sprinkler. Water introduced

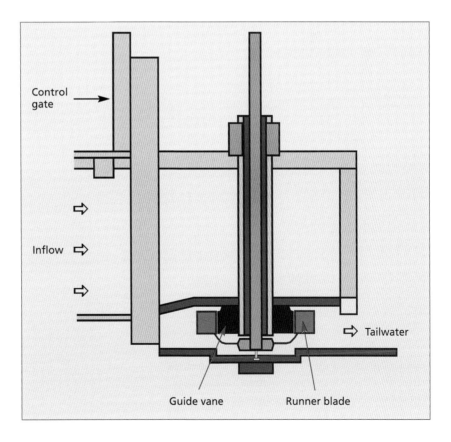

Outline diagram of a Fourneyron turbine.

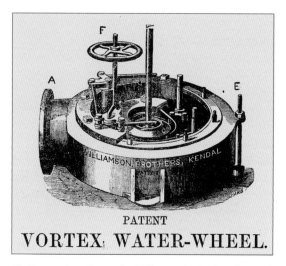

The Vortex turbine illustrated in Williamson Brothers' catalogues of the 1870s. (Museum of English Rural Life)

at pressure through a vertical pipe at the centre is diverted into two horizontal pipes. Its escape through vents at the end drives the wheel round. Dr Barker gave his name to a wheel of this type, which he designed about 1740, but it was not until a hundred years later that it was turned into a working machine, by James Whitelaw of Glasgow. He published a paper on the Barker wheel in 1833, before taking out a patent in 1839 for a design that used horizontal arms of 'S' shape where Barker had shown a straight arm. Whether through Whitelaw's paper or general circulation of knowledge, the Barker wheel was evidently sufficiently well known for there to have been one working in northern Ireland in the early 1830s. The Ordnance Survey Memoirs for 1834 on northern Ireland contains a reference to a waterwheel 'on the principle commonly called Barker's mill belonging to Messrs Mulligan in the townland of Brague'.[214] Donald & Craig of Paisley built a turbine of Whitelaw's pattern for Mr Stirrat to drive a water press. Stirrat and Whitelaw subsequently co-operated on the development of this design, which became known as the 'Scotch turbine'. They built several of these wheels, used for a wide variety of purposes in mills and on farms. Two were bought by Weymouth Waterworks Company in 1856 and 1857 for their pumping station at Sutton Poyntz, Dorset. One was employed to haul boats up the incline plane on the Chard Canal.[215]

Fourneyron's turbine was a reaction type, or, more precisely, an outward-flow horizontal reaction turbine. Water was introduced into the centre from above or below and spread out through a series of curved vanes to drive the rotor blades, which curved in the opposite direction. Fourneyron's work prompted further development of turbine technology, mainly in such parts of the world as Switzerland, Sweden, Germany, New England and France, all places where coal was in short supply.

In the British Isles, Ireland was the place without coal, but with useful streams, and an immediate interest in exploiting improvements in water power. A book published by Robert Kane in 1844, *The Industrial Resources of Ireland*, included favourable comment on Fourneyron's turbine. It prompted three men from Ulster to investigate further: William Kirk, a flax-spinner, working with Samuel Gardner, an iron founder, and, independently, William Cullen, a millwright from Armagh. They all visited France to learn from Fourneyron about his turbines. They received no co-operation from him, and returned home to design their own turbines based on what they had observed. Cullen worked in partnership with Robert MacAdam of the Soho Foundry, Belfast, to put his design into working practice. MacAdam and his brother effectively took over the design. They supplied several turbines to the flax industry of northern Ireland and, in 1870, one to replace the wheel at Catteshall Mill near Godalming, Surrey. Kirk and Gardner's design resulted in Gardner's foundry becoming one of northern Ireland's leading makers of turbines.[216]

James Thomson was a native of Belfast, who became a professor of engineering at Queen's College, Belfast, and in 1850 patented a design for a reaction turbine. The water in his turbine flowed inwards to the centre through adjustable guide vanes. He called it the Vortex. Although his turbine worked at a much lower speed than the Fourneyron, it was versatile, being capable of working with any head, from 3 feet to 300 feet; it was small, and it was 70–75 per cent efficient. These characteristics all commended it to small users, such as James Cropper

Another illustration produced for Williamson Brothers in the 1870s to show the installation of a Vortex turbine in the millstream and also the arrangement of drive to the machinery. (Museum of English Rural Life)

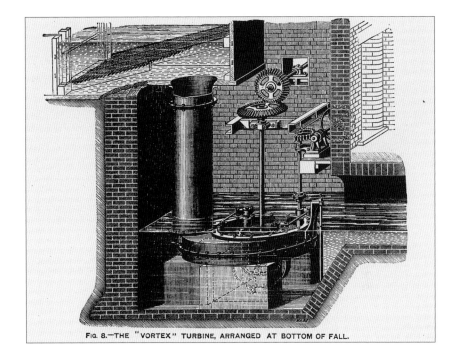

FIG. 8.—THE "VORTEX" TURBINE, ARRANGED AT BOTTOM OF FALL.

of Burnside paper mills near Kendal, who bought the first one in England from a manufacturer in Belfast. In 1856 Williamson Brothers, agricultural engineers of Kendal, gained a licence to manufacture the Vortex turbine, and made their first that year at their new Canal Head works. This 5hp turbine was supplied to a farm near Kendal, where it worked for more than a hundred years. The Vortex became the speciality of Henry and William Williamson, supplanting the chaff cutters, horse gear and other implements with which the brothers had started their business in 1853, and turning their firm into one of the major manufacturers of turbines in Britain. By 1881, when Gilbert Gilkes bought the business, 441 turbines had been made, including the one for Sir William Armstrong's pioneering hydro-electricity scheme at Cragside. Gilbert Gilkes & Co. (later Gilbert Gilkes & Gordon) concentrated further on water turbines and pumps. The agricultural implements business was run down, while production of turbines was doubled between 1881 and 1894, most of them being of higher power than the Williamsons had made.[217]

By the 1890s the water turbine was an established technology and several businesses were now producing them. Apart from the Vortex, few were of British design. Most were of European origin, such as the Giraud impulse turbine, or American, the most important of which were the Pelton wheel and the Francis turbine. James Francis of Massachusetts designed his reaction turbine in the 1850s, improving upon the original Fourneyron design by using moveable vanes to guide the water more efficiently. The Francis proved flexible in use for a wide range of situations, and was sold under a bewildering variety of brand names, including Samson, Hercules, Little Giant, Victor, American and New American.[218]

One of the the leading British manufacturers was Armfields. Joseph J. Armfield (1852–1938) joined William Munden's millwrighting business at Ringwood, Hampshire, in 1875, and bought out his partner two years later. He built up the business into one of the most successful suppliers of waterwheels and milling machinery in southern England. He started importing turbines from America in 1882, before making his own 'British Empire' turbine based upon the American designs of reaction turbine in 1887. The first of these to be sold was installed in Corfe Mullen Mill, Dorset. Up to 1949, when Armfield's turbine business was sold to the Armfield

The rotor blades of a Francis-type reaction turbine manufactured by Armfield of Ringwood, illustrated in The Engineer, *15 May 1914. (Museum of English Rural Life)*

Hydraulic Engineering Co. Ltd, 600 turbines had been sold in Britain and overseas. Most were of no more than 200hp. Many were supplied to small users in mills, farms and local industries of southern England.[219]

The impulse turbine patented by Lester A. Pelton in America in 1880 has nearly always been called the Pelton wheel because it revolved in the vertical plane on a horizontal axis, and because it allowed more air

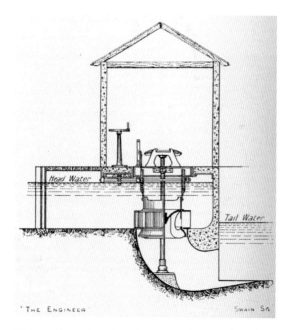

The flow of water through a turbine, sectional drawing from The Engineer, *15 May 1914. (Museum of English Rural Life)*

circulation than most turbines. Its operation is similar, however, driven by a high-pressure jet of water directed by a nozzle on to cup-shaped buckets on the rim, the pressure turning the wheel at high speed. Pelton wheels have usually been made of solid metal and enclosed in a metal casing. Although there were limits to their size and the power that they could produce, Pelton wheels became quite popular in a number of applications, including mining and some of the early hydro-electricity generators, where large falls were available. C. L. Hett of Brigg, Lincolnshire, was among the first to make Pelton wheels in Britain, his first being in 1891. Gilkes, who took on Hett's turbine business in 1895, subsequently became the major supplier of the wheels.[220]

From the late nineteenth century onwards, replacement of vertical waterwheels by turbines became a common means by which industry could continue with water power. Many corn mills followed this route – Caudwell's mill in Derbyshire was one that installed a turbine made by Hett – as did a number of farms and estates. In the 1890s an Armfield turbine was installed on the Bryanston estate in Dorset, along with the waterwheel for pumping. Robert Campbell was one property owner who enjoyed state-of-the-art technology on his estate at Buscot in Berkshire, including steam ploughing, a light railway and a processing plant for crops. He installed an Oldfield water turbine of the impulse type about 1870. It lasted until about 1920, when the turbine shaft broke and was not considered worth repairing. Paper mills, textile mills and other manufacturers also used turbines, from the small-scale to the larger mills, including New Lanark and Quarry Bank, Styal.[221]

For many users, it might be said that the water turbine proved to be a stop-gap. By the mid-twentieth century many rural corn mills had closed, and water-powered pumps on farms had been replaced by oil engines or mains electricity. It was just as true, however, that the turbine prolonged the use of water power in many works that might otherwise have turned to an alternative source when their vertical wheel needed replacing. For nearly a century the water turbine thus made an important contribution to water-powered industry. Meanwhile, turbines had

One of Charles Hett's Pelton wheels, in an engraving from The Engineer *of 1891, showing the water jet at the base. (Museum of English Rural Life)*

become established at the heart of hydro-electric power generation, with some being able to produce as much as 150,000hp compared with the 250hp of the largest waterwheels.[222]

HYDRAULIC ENGINES AND RAMS

Using the pressure of falling water to lift some of it to a greater height was the principle behind two inventions of the eighteenth century. The hydraulic engine was the first of these, developed in Germany in the mid-eighteenth century. It used water forced at high pressure in a column as a means of raising water up another column. A head of water was created in a reservoir. Some was drawn off into a pipe, and the pressure of the column above it built up to operate the piston of a cylinder, from which drive to a pump could be taken.

The hydraulic engine was introduced to Britain during the late eighteenth century, with the first example being installed at Allenheads lead mines in the Pennines. John Smeaton, Richard Trevithick and Sir William Armstrong all made improvements to hydraulic engines, which were used a great deal in mining during the first half of the nineteenth century, often in combination with waterwheels. Armstrong developed his hydraulic engines further for such uses as passenger lifts, dock cranes and bridges. Tower Bridge in London is the best-known example of a bridge originally operated by hydraulic power.[223]

Another self-lifting water pump was the hydraulic ram, developed in the late eighteenth century. It was invented independently by John Whitehurst in England and Joseph Montgolfier, of hot-air balloon fame, in France. Montgolfier made the practical models in 1793, licensing his patent to Boulton &

A Pelton wheel of 20 feet in diameter supplied by Gilbert Gilkes & Co. Ltd to a tinplate-rolling mill at Cwm Avon near Port Talbot, one of two installed in 1903 to replace four overshot wheels. (Museum of English Rural Life)

Developing the Technology of Water Power

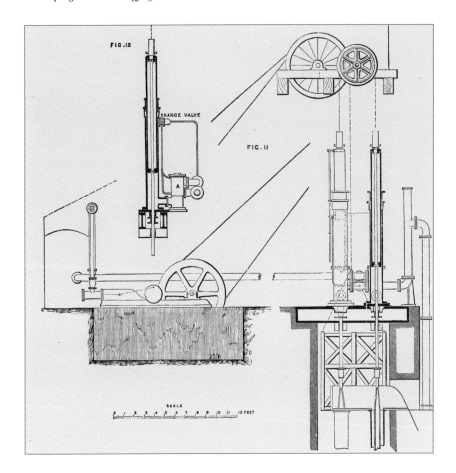

A hydraulic engine for winding at a lead mine near Richmond, Yorkshire. (The Engineer, v. 49, 1880, p. 318.) (Museum of English Rural Life)

Watt in 1797. It was not until after the Napoleonic Wars that the machine was used seriously in Britain, and from the 1820s onwards it was supplied by a number of manufacturers.

The hydraulic ram is a momentum machine in which water descending through a pipe into the ram is forced by its own pressure into another delivery pipe, which takes it up again to the higher point where it is to be used.

The main use of the hydraulic ram was in water supply, and it was sold principally in the days before mains water for farms and country estates. According

In the 1950 catalogue issued by John Blake Ltd, their Hydram hydraulic ram is shown sited in the millstream lifting water to the houses on the valley side. (Museum of English Rural Life)

to the catalogue issued about 1950 by Blake's of Accrington, one of the makers of these pumps, there was hardly a member of the aristocracy whose estate did not have one. One of their rams was also pumping water for the Taj Mahal in India. Rams were small and relatively cheap, and could be fitted into a modest stream with a low fall. Damming the stream, therefore, did not need to be on a large scale. The only other engineering needed was the long lengths of pipe-work bringing the water to the ram and taking it out again, far less than required for pumps driven by vertical waterwheels or by turbines.[224]

NEW BUILDINGS

The mills of the Industrial Revolution were more often purpose-built than conversions of existing mills, their scale of operation being so much greater. Their builders had a number of considerations to take into account. They needed space for the spinning frames, mules and looms, in terms of floor area and height, and they needed strength in the upper floors to support all that weight. Making the building fire-proof became a greater concern, and this was usually achieved by placing brick or iron cladding around timber that might formerly have been exposed. They needed adequate light, too, although Arkwright compromised on that in his first mill at Cromford, which had a forbidding road frontage, prompted by his desire for secrecy about his operations. There was also a need to accommodate the line-shaft running the length of the workshops, from which drive was taken to the machines.[225]

Not least, of course, there was the provision of the water power. As well as a building a hundred feet or more in length, the site also had to accommodate the water channels. The builder and millwright had a number of choices to make over the position of the waterwheel. Would it be outside or inside the building? If outside, on the long or short elevation? If inside, at the end or in the middle of the building, or in a wheel-house attached to the main mill building?[226] Most mills had enclosed wheels, either inside the main building or in an attached wheel-house. It certainly added to the impressiveness of the power – descending to the ground floor or basement to reach the wheel was rather like entering the engine room of a ship. To the observer outside, a mill with an internal wheel is less obviously water-powered, especially with surviving examples where the pond might be filled, and extensions for steam power obscure its origins.

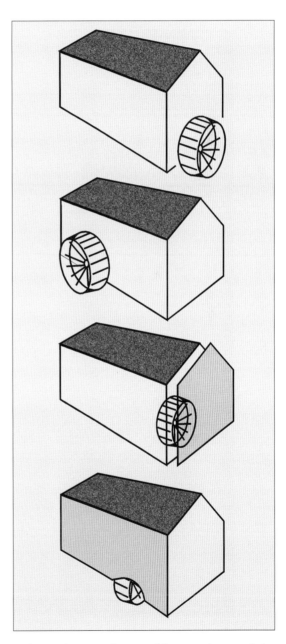

A schematic diagram to show some of the principal ways in which wheels were positioned in the mill.

Developing the Technology of Water Power

The fulling mill at Helmshore, Lancashire, built in the late eighteenth century.

The buildings on the whole were utilitarian in style. The multi-storeyed mills were often timber-framed, until concern for fire-proofing introduced greater use of iron. They were usually built in the local vernacular, such as the solid stone of Lancashire's Higher Mill. Despite its large scale, Quarry Bank Mill was one building that retained broadly domestic features, similar to the weavers' cottages in the village, especially in the design of its windows. Occasionally, designers allowed themselves additional styling. Arkwright's Masson Mill of 1783 was given greater embellishment, as befitted the factory of an industrialist of the first rank. Palladian architectural principles inspired Venetian-style windows and a central cupola. Elegant pedimented frontages were given to several water-powered industrial mills built up to the 1830s, such as the Whitchurch silk mill or Wye Mill at Cressbrook, Derbyshire, which managed to look rather like a country mansion. The mills at New Lanark also followed this style.[227]

A more ornate style was sometimes applied to corn mills, usually those centrally situated on the estate in whose lands they lay. James Paine built a mill at Chatsworth, Derbyshire, in the eighteenth century designed to be a feature of the park as well as for grinding grain. Although badly damaged by a storm in 1962, the building still stands. Howsham Mill in Yorkshire was built in Gothic style by John Carr of York in the mid-eighteenth century. Kings Mills at Castle Donnington, on the River Trent, were built by the Hastings estate in the 1790s, also in Gothic style to match nearby Donnington Hall. The mills were used to grind flint and alabaster.[228]

What might be regarded as the characteristic water corn mill was a product mainly of the eighteenth and nineteenth centuries as mills were built and rebuilt to increase production for a growing population and economy. The result was a large building of many floors, often with projecting lucam for hoisting sacks to the top floor. The scale was largely determined by the need to accommodate

The Masson Mills outside Cromford, built for Sir Richard Arkwright. The section with the tower was an extension added to the original part of the building with its Venetian windows.

Elegance in the external finish of Stanley Mill, Gloucestershire, with stone quoins, Venetian windows capped with stone, and, on the top floor, rubbed brick arches.

Developing the Technology of Water Power

The flour mill at Chatsworth was built just across the river from the house, and was in keeping with its style. A leat ran underground fed from a weir across the Derwent behind the photographer.

Piccotts End Mill, Hertfordshire, in weather-boarded external finish, with the external sack hoist housed in the projecting lucam.

greater numbers of stones, dressers and other ancillary machines, and especially for the provision of storage for grain and flour. Storage bins for grain storage were on the upper floors, and a gravity flow of processes took it down through the milling floors below. Mills engaged in long-distance trade supplying the major cities were among the largest. Most were of brick or stone, but in the eastern counties white-painted weather-boarded structures were predominant. Cast-iron framework was introduced into the construction of mill buildings by the late eighteenth century, even though most of the mill gearing might still be of wood.

9 Flour-Milling in the Industrial Economy

Alongside the new industries driving forward water-power technology, the milling of flour continued to be a major user of this particular source of energy. The population was increasing rapidly during the eighteenth and nineteenth centuries, and with it the demand for flour. When the first census was taken, in 1801, the population of Great Britain was recorded as 10.6 million, which was about 3 million more than in the middle of the eighteenth century, and perhaps 4.5 million more than at the beginning. By 1851 the British population was almost double what it had been in 1801.[229] At the same time the proportion of the population of England and Wales living in towns was expanding, reaching 50 per cent in 1851. There was also growing agricultural demand for grain ground as animal feed.

In order to meet this growing demand, greater milling capacity was needed. This meant more mills, bigger mills and more power, and the flour-milling industry from the mid-eighteenth to the mid-nineteenth century got all three. Water power was by no means the only contributor: additional power came from steam and wind. Grain-milling was one of the first industries to use steam power, introduced in the 1780s, almost as soon as Watt had developed rotary drive. The first steam mills were not especially successful, but by the 1850s they could be found in many market towns. At this stage, however, steam power's contribution to the industry was still relatively modest.

Of greater importance for the expansion of capacity before 1850 was the windmill. Most of the completely new corn mills built in the eighteenth and early nineteenth centuries, before the large steam mills achieved dominance, were windmills. In lowland Britain, where most of the cereals were grown, windmills were built in large numbers; there were about 150 in Kent, for example. Suffolk, where previously watermills had almost had a monopoly, now gained many windmills, and Lincolnshire was another great county for windmills, in the villages and around the towns. Indeed, throughout the corn-growing eastern counties, one of the characteristic sights became the tall tower windmill, usually with the most up-to-date, efficient sails and other modern equipment. These tower mills were still being built in eastern England into the 1870s. Some, such as the tower mill at Legbourne, Lincolnshire, were built alongside watermills, where the miller had evidently decided that the limits of expansion of water power had been reached. Having a windmill as auxiliary power was quite a common alternative to rebuilding the watermill. It could be cheaper to build and avoided problems over rights to use water. It could also offer a useful combination of power: when there was insufficient (or too much) wind, the watermill might be working, and vice versa. There were at least nine examples in Essex at the end of the eighteenth century. Although windmills were continually being edged out by urban expansion, they were important for supplying some of the large towns in eastern England. Newcastle upon Tyne had one of the greatest concentrations of windmills before steam power became established. In 1822 there were thirty-seven windmills grinding 1,100 quarters of wheat a week, but only six watermills grinding 270 quarters.[230]

So effective were the new windmills in some of the lowland counties that by the mid-nineteenth century they were edging out the watermills, which were starting to decline in numbers. This state of affairs was confined to those areas where the avail-

One of Dorset's water-powered flour mills, at Sturminster Newton. The building dates from the seventeenth to eighteenth centuries. Its waterwheel was replaced by an Armfield turbine, which helped keep it at work into the early 1950s, when this photograph was taken. (Museum of English Rural Life)

ability of water power was more limited, for example, Lincolnshire. Where there was a more abundant water source, numbers of watermills were maintained, and there were fewer windmills. Derbyshire was an example of a county where this was the case. In southern counties, away from the main grain-growing districts, there were far fewer new windmills. In Dorset and Buckinghamshire, for example, water power was evidently sufficient, at least for local needs.

Despite the influx of new windmills and steam mills, the watermill still had an important role to play in meeting the demand for flour and meal, but not universally. There were many places where water-power sites were worth far more as textile mills than as corn mills, and many were converted from the latter use to the former. This development was helped by the fact that the new water-powered industries tended to be in the northern and upland regions, where the growing of cereals was less important, and thus the local demand for corn-milling was reduced. In the southern counties there was much greater demand for water corn mills. This geographical distribution is perhaps one reason why the word 'watermill' generally conjures up a south-of-England corn mill in the popular imagination.

Where watermills were important in corn-grinding, their productivity was increased during the period of the Industrial Revolution. This was done in four main ways: new mills were built; some existing mills were made to work harder; other mills were rebuilt and extended; and some were converted from other uses to grain-milling.

The number of completely new mills was limited, simply because finding a site for one was not an easy task. There were few unused sites, and there were other businesses competing for them. By the end of the eighteenth century almost every stream, even the smallest, in the grain-growing regions of lowland Britain was clogged with mills. This was especially true of the area within ready reach of London. The Thames itself had a series of mills above the city to Abingdon and beyond, and its tributary rivers and streams all had their fill of watermills. On the Kennet, the open stretch of river between Newbury and Reading had a mill approximately every 2 miles, while within the environs of these two towns the mills were more closely spaced. The Loddon, a modest river of a little over 10 miles, contained no fewer than five mills. There were four mills within 1 mile, all grinding corn, on a tributary stream of the River Len at Hollingbourne in Kent. The River Wandle in Surrey supported thirty-eight mills, all at work in the 1860s, although not all used for grinding flour. Their combined power output was reckoned to be 781hp, approximately 20hp each.[231]

Despite the shortage of sites created by the many competing interests for water power, some new watermills for grain-milling were built in the late eighteenth and early nineteenth centuries. Danby Mill in north Yorkshire was built about 1800, space being found along the River Esk by taking the water almost directly from the stream to drive an undershot wheel. Some opportunities for new mills were found even in the eastern counties. After the Chelmer Navigation in Essex was opened, a number of new mills were built along that river and the Blackwater to take advantage of the improved transport. Navigation Mill at Warwick was constructed to use surplus water from the Warwick & Napton Canal.[232] In common with most new watermills, this one was large – it had five pairs of stones – and was designed to make the most of the site and the latest milling technology.

Milling capacity could also be expanded by making more intensive use of existing mills. Mills that had worked part-time for local customers could become full-time, supplying distant urban markets. Small mills where the miller was also a farmer or engaged in another trade were most likely to be affected. Fairbourne Mill in Kent was one example: between the 1760s and 1780s its output increased a hundred-fold, without any rebuilding or other additions to the capacity of the mill.[233] Increasing the working of a mill might be restricted by the availability of water, however, and this could affect even large mills. At Canterbury there was a mill that could grind 500 quarters of wheat in winter, but only 300 in summer because there was insufficient water. In Sheffield a mill that could work six pairs of stones in winter could only manage one pair in summer, and that for no more than four hours in the day.[234]

The third, and far more common way to increase milling capacity was to rebuild existing mills and enlarge them. There were two objects: first, to increase the power from the waterwheel and so add extra pairs of stones, and second, to add more storage space. On the Chelmer and Blackwater in Essex, Beeleigh mill was rebuilt in the 1790s and Barnes mill about 1800. The capacity of Little Baddow mill was increased from 35 to 150 quarters per week in about 1811, while Hoe mills went from

Daniel's Mill, near Bridgnorth, Shropshire, with its mid-nineteenth-century iron wheel designed to work either as pitch-back fed from the high tank or as breast-shot.

Flour-Milling in the Industrial Economy

The engineers Whitmore & Binyon of Wickham Market supplied equipment for mills, including waterwheels, as illustrated in their catalogue of 1876. (Museum of English Rural Life)

three to eight pairs of stones. Abbots Mill, Canterbury, was rebuilt in 1791. Its new building had six storeys and floor dimensions of 72 feet × 52 feet 6 inches. With eight pairs of stones and fourteen dressing mills, the rebuilt mill could grind more than 500 quarters of grain a week. The rebuilding cost £8,000 at a time when the insured value of most water corn mills was no more than £1,000. Indeed, the low valuations of most watermills kept the average values for England and Wales low. In 1751–5 that average was £315. As more mills, such as Canterbury or Little Baddow, were built or rebuilt, so the average value rose to £817 by 1796–1800 and £1,042 by 1816–20. On a more modest scale the watermill at Norwell in Nottinghamshire had its water supply increased at the time of enclosure in 1832, to give greater power, while additional storage and other outbuildings were added to the site as business was growing at this time. Some new work to expand production came quite late. One of the mills still surviving, Daniel's Mill near Bridgnorth, underwent some major work in 1854, with a new wheel of 40 feet in diameter, and extensive work to drive the wheel in pitch-back form, drawing water from a reservoir some 300 yards away.[235]

Some of the largest mills were built or rebuilt in the 1780s to supply the growing industrial towns of north-west England. With competition for water power from the textile mills leaving many of the Lancashire cotton towns with very few corn millers, more intensive use was made of other sites, most of them further south. The Mersey Mills, Warrington, was probably the largest water corn mill to be built, with twenty-one pairs of stones. The Dee Mills at Chester, fed from a weir that dated back to Norman times, were rebuilt and enlarged in the late eighteenth and early nineteenth centuries until they had eighteen pairs of stones driven by three external wheels. Frodsham Mill, with fourteen pairs of stones, was on a similar scale.[236]

Some of the mills used for other industries were converted to corn-grinding. This was important in some localities. While the growth of the textile industries in the north competed for mill sites, in the south the opposite was true. The decline of wool manufacture meant that many mills were now more valuable for grinding corn. In Gloucestershire at least fifty former textile mills became corn mills. Several of the mills of the Kennet valley that in the eighteenth and nineteenth centuries were engaged in the flour trade to London had had a previous life as woollen mills. Botley Mills in Hampshire had been used as a paper mill for several years, but was converted to corn-grinding in the 1840s. The decline of the Wealden iron industry left many old forge mills available, some of which were converted to corn. In Sussex, Conster Mill was built in the early nineteenth century on the site of an iron forge on the River Tillingham, and the mill built at Robertsbridge in about the 1790s used another former furnace site. Stream Mill, near Chiddingly, a site recorded in Domesday Book, was converted from iron-working to corn-milling some time in the eighteenth century. In the Midlands, as the iron industry moved to working on a larger scale in steam-powered works, mills were freed for conversion to corn-milling. Eight or nine mills in the Tame Valley changed from working iron to milling corn between 1800 and 1829. All of them continued in work grinding corn up to the late nineteenth century. For a number of these mills this was a reversion to their former use; many had been converted to iron-working in the seventeenth century.[237]

There were losses of watermilling capacity as well. Besides the sites converted to textile mills, a few mills were already being lost to urban expansion by the early nineteenth century. The construction of canals also had an effect. The canal companies were anxious to secure their supplies of water, yet were constrained from drawing large quantities from rivers by the rights of mills. A number of Acts authorizing canals set limits on the amount the companies could draw, especially in northern areas where powerful industrial interests had significant influence. The flour millers in the south had less clout, and there were several instances of canals reducing the water supply to mills. The Chelmer-Blackwater navigation in Essex caused at least two mills to lose as much as a third of their output. Some canal companies, the Ouse Navigation among them, bought out the mills to prevent litigation from aggrieved millers. Those mills might then be leased, or simply closed. The windmill at Wilton in Wiltshire was built in 1820 as a replacement for some of the watermills closed as a result of the construction of the Kennet & Avon Canal.[238]

The productivity of corn mills was improved with the application of the new developments in mill technology and millwrighting. John Rennie noted in the 1780s that corn mills more than most were in need of such improvements, and during the following decades there was widespread upgrading

Four pairs of stones at Longbridge Mill, Hampshire.

of watermills. Rebuilding and enlargement of mills was likely to include replacing wheels and gearing with better, more efficient equipment. Sometimes undershot wheels were placed by more powerful breast-shot or overshot wheels, and wheels of iron or hybrid construction were introduced in place of wood. Opportunities for changing the type of wheel in a mill on a crowded river with low fall could be limited, but a number of mills did manage to do this. The mill at Alvingham in Lincolnshire had a breast-shot wheel installed when it was extended in the 1780s. At Haxted Mill, Edenbridge, an undershot wheel was replaced by an overshot wheel, work which involved damming the River Eden to raise the level behind the wheel by 10 feet.[239] The expense of deepening wheel-pits and enlarging water channels to maximize the fall could be considerable. And, having done the work, the miller might incur the wrath of another miller claiming his mill had been adversely affected, or from a landowner or farmer claiming that water meadows had been drained. It was not uncommon for a certain amount of wrangling to ensue.

Haxted Mill was fortunate in securing sufficient fall for an overshot wheel. It was difficult to do this throughout much of southern and eastern England, and by the mid-eighteenth century nearly all the sites in these areas where an overshot wheel could be installed probably already had one. The rivers in East Anglia had insufficient fall, however well engineered the artificial watercourses might have been. The average fall on the River Stour in Suffolk, for example, was 6 feet 3 inches, and 10 feet was reckoned to be the minimum needed for an overshot wheel. It was a similar story on most other rivers in the county. There were greater opportunities for installing a breast-shot wheel, for which a fall of between 4 and 10 feet was needed; the efficiency of the new breast-shot wheels meant that the many mills that used them saw a gain in productivity. Nevertheless, the shortage of water power did mean that the lowland zone of the country, the greatest producer of corn, faced the greatest difficulties in expanding the output of existing mills. The pace of change was slower, therefore, than many might have wished, with examples of undershot wheels still being replaced late in the nineteenth century.[240]

The introduction of waterwheels made wholly or partly of cast iron often had to wait until the second half of the nineteenth century. Foundries local to the mill might not have the capacity to make a wheel of sufficient quality and, until the railway network had expanded, it was often too expensive and time-consuming to fetch iron wheels from a distance. The

One of Norfolk's grand mills, at Buxton. (Museum of English Rural Life)

A diagram produced by John Farey showing the working of a Savery steam pump to pump water from the stream back up to the waterwheel. The boiler at B produced steam which, when admitted into the cylinder at A, condensed and caused water to be drawn up from the stream at H, and deposited into the reservoir at R.

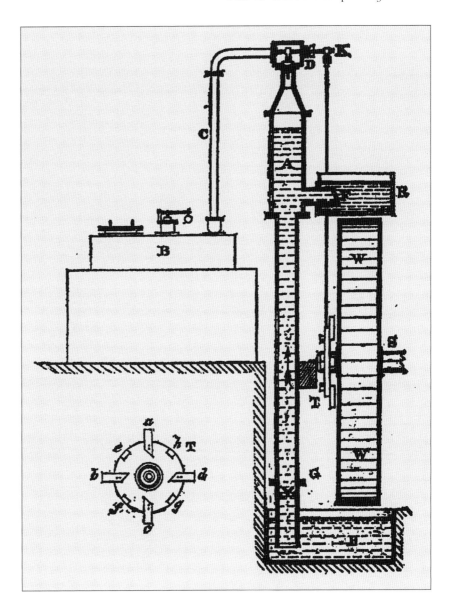

otherwise support only an undershot wheel. Sometimes the steam engine was used to extract additional supplies of water for the wheel from deeper underground.

Smeaton used the combination of steam engine and waterwheel in the closed cycle on a number of occasions, including Long Benton and Walker collieries in the north-east. Many other engineers employed steam engines in the same way. Arkwright bought his first steam engine from Boulton & Watt, possibly in 1780, and used it to feed the waterwheel at his Wirksworth mill. The Shudehill Mill at Manchester, where the steam engine was definitely acquired in 1783, also opted for this arrangement. While cotton masters remained concerned about the uneven vibrations from a steam engine, using steam to pump water to the wheel was a reliable proposition.

Watt patented his rotary drive in 1781. He had actually been beaten in the race by Matthew Worsborough and James Pickard, who had designed a rotative engine in 1779 and installed it at Snow Hill rolling mill, Birmingham, where water power was in short supply. Boulton & Watt, however, had the stronger marketing muscle, and from the first installation of Watt's rotary engine in a cotton-spinning mill at Robinson's mill, Papplewick, Nottinghamshire, in 1785, theirs was the more common type. The expiry of James Watt's patents opened the way for the development of steam engines running at higher pressures and for the entry of new firms making them. New boiler designs were developed – the 'Lancashire' and 'Cornish' types – and these enabled the production of steam engines that were bigger, more powerful and more efficient. By the 1820s the steam engine was ceasing to be an add-on, which proved useful when the water was low, but was in general contention as main power. Factory owners were no longer confined to a steam engine that could generate no more power than a waterwheel. Now, they could buy a bigger engine that was capable of driving large numbers of machines. Steam was becoming the serious proposition for new mills, and as the nineteenth century progressed it was increasingly the best option for replacement power too.

The steam engine offered the potential for unlimited power, and it was expansion of mills beyond the capacity of their waterwheels that often proved the tipping point in the adoption of steam. Most mills using the Arkwright spinning frame had between 1,000 and 2,000 spindles, for which a waterwheel of about 20hp was quite adequate; this was what most water-power sites could support. Quarry Bank Mill, Styal, was unusual in having the water-power capacity to expand well beyond that, but when further expansion was required, with the addition of a weaving shed, the limit of its waterwheel had been reached. Sourcing sufficient power for the weaving would have entailed considerable capital expenditure, either on enlarging the reservoir and leats, and probably installing a new wheel, or on an engine and boiler house. In the event the new weaving shed was powered by steam while the waterwheel continued to drive the spindles. The experience of the woollen industry was similar. Even with five wheels at the big King's Stanley mills in Gloucestershire, further expansion in 1822 necessitated the purchase of a steam engine of 40hp.

The desire for larger mills brought into sharper balance the capital costs of steam and water. In his treatise on mills in 1861, Fairbairn made a comparison of the costs of working his waterwheels at Catrine and steam power in the same district, and came out with little to choose between them. In his opinion, the consequence was that 'in most cases where a large amount of power is required, the chief source from which it must be derived is steam'.[256]

New types of textile machinery were likely to need more power, and they were often more complex. Their introduction was often the occasion for the adoption of steam power. The fixed costs of water power increased with the installation of such machines, while technical development in its engineering made steam power more attractive to work the new machinery. Jenkins and Ponting observed of the woollen industry that it was 'no coincidence that the development of power loom weaving in woollens coincided with steam power improvements'.[257] In the cotton industry a second phase of mechanization began in the 1780s with the introduction of the spinning mule invented by Samuel Crompton. This could produce finer thread than either Arkwright's water frame or the spinning jenny of James Hargreaves, and quickly established itself in the industry. By 1810, mule spindles probably far outnumbered the other types. At first its drawback was that it did not seem to lend itself easily to mechanical power. Initially it was worked by hand and attempts in the 1790s to devise gearing from waterwheels appear to have had limited success. By the time the mule was mechanized, after 1800, it was more likely to be powered by steam, and the mule-spinning mills were tending to congregate in Manchester and the other urban centres. This represented the first major break with water-powered mechanization of the cotton industry. With many of the best sites for water power occupied by mills running Arkwright machines, and with steam engines now capable of meeting the demand for power, new mills to operate the mules were likely to be built in towns and to be powered by steam.[258]

The same developments occurred with cotton-weaving. The power loom was not successfully introduced until the second quarter of the nineteenth century, by which time the factory owner was likely to choose steam power, as was the case at Quarry Bank.

THE QUESTION OF COST

To the businessmen who ran the mills, the relative cost of water and steam power was as important as the technical matters of reliability. There were two elements to this: the capital costs of buying and installing the waterwheel, weirs, leats and reservoirs compared with those of a steam engine, and the running costs. In the 1780s, when Watt's first rotative engines were being produced, water was more than competitive on both counts. During the following decades the capital cost of water power rose, while that of steam declined. At some point, around 1830, the two were more or less at parity, and thereafter steam drew ahead. The running costs of steam had also fallen considerably by the 1830s, but water power was still able to compete.

Often the wheel represented the least of the costs for the user of water power. Wooden wheels were especially cheap and, at the low end of the range, a basic wheel of perhaps 10hp could be got for as little as £30 at the end of the eighteenth century. Most wooden wheels at this time were valued at less than £300 (although considerably more was spent at some of the larger cotton mills, such as the West Mill at Belper, where the new wheel mainly of wood installed in 1795 cost nearly £630). Iron was more expensive, and a more efficient breast-shot suspension wheel was disproportionately even more costly, so that a mill business could reach a point of diminishing returns. Even so, a standard iron wheel of 10–20hp was affordable in the mid-nineteenth century for most mill owners, at £80–90. Bigger wheels in the new factories cost more, sometimes considerably more. Haden Brothers of Trowbridge installed a wheel 16 feet diameter and 11 feet wide at Eastington, Gloucestershire, in 1820 for £250. In 1832 a new wheel for a cotton mill in Bollington, Cheshire, was supplied by James Lillie of Manchester for £690. The scale of cost rose until at Eagley Brook the wheel of 140hp in 1828–9 cost £4,800. These figures do not always include the costs of installation and construction of the wheel-pit. A wheel-house at Belper West Mill added another £323 to the total bill.[259]

It was the cost of constructing the dams, reservoirs and watercourses that rose far more quickly and added substantially to the capital sums needed for water power. At the Bollington mill in Cheshire, in addition to the new wheel, the building of a culvert more than 100 feet long cost £813. On a larger scale, in the 1780s George Robinson constructed reservoirs extending to 30 acres and leats to serve his five mills at Papplewick. The cost was £3,720, almost twice that of the wheels. These somewhat extreme measures were designed to compensate for a lack of fall on the sluggish stretch of river where Robinson was based. He could, perhaps, have abandoned Nottinghamshire and moved to somewhere with better water power, as sites were still available at this time. William Douglas and Joseph Thackeray were two mill owners who were prompted by inadequate water power to move from their mills near Manchester, Douglas to Holywell, north Wales, and Thackeray to Cark in Cartmel. Robinson instead decided to augment his power with the Boulton & Watt engine. As sites for water power decreased in number, the costs of constructing and enlarging reservoirs and leats increased disproportionately. They were lower on established sites, especially if expansionist aims were modest. This was one way in which water power remained competitive in the West Country, where ponds and leats might be enlarged but little by way of new construction was needed. By the 1830s, however, the costs of installing the power supply were telling against water power. Reckoned as cost per horsepower, the capital for new mills of the 1780s, such as Douglas's at Holywell, could be as low as £15. By 1830, Quarry Bank Mill's valuation gave the equivalent figure as £75 per horsepower; at the Catrine mills in Scotland it was £77, and at Arkwright's Bakewell mill, £93.[260]

While the cost of installing water power was rising, the cost of a new steam engine was falling. The steam engine of the late eighteenth century was costly to buy, and its fuel consumption high. Engines with rotative drive bought from Boulton & Watt in the late 1790s to early 1800s were priced at £1,000 or more for engines of 20–30hp. The firm, for example, quoted £1,497 for an engine of 32hp for a Scottish paper mill in 1805. Other builders, such as Fenton, Murray & Wood of Leeds, sold their engines for similar prices. Engines of lower power were cheaper – about £500 for one of 10hp, the equivalent of many a corn mill's waterwheel – but the price per horsepower was higher for the smaller engines. This higher cost per horsepower for lower-powered engines certainly deterred many potential customers, in the woollen mills of the west of England, for example, where power requirements were generally low. As well as buying the engine, there were a number of costs to be met for delivery and installation of the framework in which the engine would sit, and the pipes and drives from it. At existing mill sites it was unlikely that the engine would fit into the building, so there was the cost of building work for the engine house that the large engine and its boiler would require. Every site and installation was different, of course, but it was not unusual for these ancillary costs to add £300 or more to the total bill.[261]

By the 1830s the steam engine cost less to buy, and offered improved reliability. Most cotton masters were more prepared to accept that it would run sufficiently smoothly not to break the thread, making it a more serious option.

When it came to running costs, the cost of fuel was the major item for the user of steam power. The long-term trend in the price of coal was downwards. The average price in London in the 1830s, at 20s to 23s a chaldron, was about half what it had been in the 1800s.[262] Although the price trend was downward, there were considerable variations between different parts of the country, determined almost entirely by the cost of transport from the pit to the user. Before the spread of the railway network reduced the differentials in price, the relative cost of coal was a serious consideration in the decision between water and steam power for the mill. Away from mining areas and major towns, coal cost far more, and its price fluctuated less. The high price of 11s–12s a ton for coal was a strong influence on George Robinson's decision to keep his waterwheels running at Papplewick in the 1780s. Distances from the pit did not have to be great for there to be a significant difference in the price of coal. Many Pennine mills were not far from the coalfields but, before canals and railways penetrated those hills, transport was difficult, and the price in these districts was much higher than in Leeds or Bradford.

On the regional scale the price of coal became a major influence on the location of the textile industries. Throughout most of East Anglia, the south Midland and southern counties of England, the price of coal was more than twice that in Yorkshire, while in the Midlands and Yorkshire outside of the coal-mining district it was about one and a half times. As steam power gained its competitive edge, the textile industries in the southern counties grew weaker. In 1830 a report for a Select Committee of the House of Commons noted that the woollen industry had 'in great measure migrated from Essex, Suffolk and other southern counties to the northern districts where coal for the use of the steam-engine is much cheaper'.[263]

Transport, of course, was the key, both because of the distances involved and because of the more limited development of canals in many parts of southern England. Boulton & Watt's first sales in the west of England were to Bristol and Bath, places to which coal could be transported with relative ease. The coming of the railways did not immediately change that balance, for the network grew most quickly with links from the mining districts to major towns and cities, and within the coalfields themselves. It took some time for branch lines to penetrate such counties as Wiltshire and Gloucestershire.[264]

A lower cost of coal in the West Riding resulting from the proximity of the mines, and good transport facilities with the development of the canal network, stimulated the growth in the use of steam power in the main towns of the woollen and worsted districts of Yorkshire. There were limited opportunities for expansion of watermills, which were having to be built in more isolated situations along the Pennine streams, and the problems of intermittent water supply were at least as great as in other districts. All of these factors were prompting mill owners to follow John Marshall and look seriously at steam power before 1800. By 1830 their choice in favour of steam was already decisive.

ADHERENCE TO WATER POWER

Water power had one big advantage over steam and that was very low running costs. There was rent to pay to landowners for the right to extract water, but otherwise the fuel was free. Waterwheels and transmission required little maintenance; ponds and watercourses needed regular cleaning and the soundness of dams had to be maintained, but these were not usually heavy expenses. Depreciation to cater for replacement of the wheel and major works to dams and leats was the major item to take into account. Many manufacturers accounted for it at about 7 per cent. Low running costs meant that, once the investment had been made in water power, it could be cost-effective, even in competition with steam power, the cost of which was falling. It could still be worth investing in new water-power installations and this was one of the reasons why water power continued in use through the nineteenth century and beyond.

According to the factory return of 1870, 55,620hp used in manufacturing industry came from water power. There was substantial under-recording and John Kanefsky suggests that the total may have been nearer 100,000hp, adding that even this might be a conservative estimate. The census of production for 1907 gave a figure of 112,429hp as the capacity of water power in manufacturing; allowing 70 per cent of capacity as being in use to make the figure comparable with the 1870 return suggests that 80,000hp were still being produced from water. In addition, there was the water power still generated in mining, agriculture and other sectors.[265]

Mill owners were sensitive to the costs of power, and made detailed comparisons between water and steam. The manager of Quarry Bank Mill was regularly carrying out such calculations. In 1849 he concluded that the running cost of steam power equivalent to the 100hp of the waterwheel would be £274 a year greater; in 1856 he valued the water power at about £280 a year. In terms of cost per horsepower, water and steam were closely comparable in the 1830s. Water power based on Quarry Bank worked out at £18 3s 4d per horsepower, steam engines in Manchester at £17 11s 6d. What told in favour of the waterwheel was the high cost of coal carried to Styal, which was in a rural location, even though it was not far from Manchester.[266]

The low running costs of water power meant that, until the 1830s–40s at least, the steam engine did not automatically introduce savings into textile manufacture. There thus remained many incentives to retain water power, and even to continue investing in it. The mill owner who had recently made a heavy investment in a new wheel and the waterworks on a site that was ideal for water power was likely to want to continue with it. This was true at Quarry Bank, and also for the owners of mills at Belper, Macclesfield and Stockport. The Derwent Valley mills of Derbyshire continued to use water power, several of them into the twentieth century, long after many other areas had transferred to steam. While Marshall paved the way with his steam-powered flax mill in Leeds, the firm of Colbeck, Ellis & Willis, with abundant water resources in the Washburn valley at Fewston, were among the rival firms able still to provide effective competition using water power. Even within Yorkshire, the geography of change was very localized. Water power remained more important in the Pennine districts around Halifax, Huddersfield and Saddleworth, for example, where the cost of bringing the coal up from the pits reduced the economic effectiveness of steam. As well as maintaining existing mills, there was more likelihood of new wheels being installed in these districts, while around Leeds the use of steam power grew and waterwheels were taken out of use.[267]

Even when a coal mine was almost on the doorstep, consideration of the amount of capital invested in the waterwheel and watercourse could make it cheaper to continue with water power. Mill owners certainly were slow to replace their waterwheels entirely. Many did accept steam as auxiliary power to keep the waterwheel turning, to tide them over dry seasons or as a means of expanding the mill beyond the capacity of the existing waterwheels. Most of Boulton & Watt's first orders from textile mills were for engines to be used in this capacity.[268]

The relatively low cost of water power made it worth continuing to invest in it and new wheels continued to be installed between the 1820s and 1840s. Many, it is true, were ordered by mill owners already committed to water power, and in areas blessed with water resources or particularly poorly supplied with coal. The Ashworths enlarged their New Eagley Mill in 1822–5, including a new wheel of 45hp, and, when they bought the mill at Egerton in 1828, committed themselves to the large wheel already under construction, although they soon had to get auxiliary steam engines. The Bolton area, where the Ashworth mills were situated, maintained greater dependence on water power while most of the Lancashire cotton industry was turning to steam.[269] The Fielden brothers built large new reservoirs and new wheels for their mills at Lumbutts, near Todmorden in 1830.[270] Many of William Fairbairn's greatest examples of mill engineering were also built at this time.

The Ashworths were among those who calculated the costs of water against steam power. They reckoned they saved £4 a week in coal by using water at their New Eagley Mill in the 1830s, while at the Egerton Mill in 1841 the saving was, they said, £20 a week. Having spent £4,800 on the new wheel it made sense to continue to run it, but it was not without cost, and some of the calculations began to look as though they might have been more to do with self-justification. At 7 per cent, depreciation on the new wheel represented £336 a year. The rent of land, waterfall and reservoirs was high: at New Eagley it came to £167 8s 6d a year. In addition, there were difficulties in maintaining a regular water supply. This combination of issues meant that the balance in favour of water was, according to the historian of the Ashworth enterprise, marginal. By the 1830s, water power was for them 'a decreasing asset, and the Egerton wheel became more of an ornamental showpiece than an economic triumph'.[271]

This begins to make it seem as if some of the calculations of cost were to demonstrate that a commitment already made to water power could still be justified. Indeed, it could, on the grounds of low running costs and the fact that the upheaval of replacing it would render an existing factory unprofitable. The same reasoning could justify expenditure on maintaining water power, and upgrading it with the purchase of a turbine. However, by 1850 steam could be reckoned to have the cost advantage for almost all major investment in new works or expansion to factories.

There was a further cost borne by water power, which was not generally quantified, and that was the cost of isolation. The rural mills, such as Quarry Bank, had the significant overhead of maintaining a factory village in which their workers lived. By the 1830s mill owners were coming to prefer to cluster together in Manchester and use a steam engine.

THE DECLINE OF WATER POWER IN THE TEXTILE INDUSTRY

By the 1830s steam was making a serious contribution to the power requirements of the textile industries. From this time until the 1870s some statistical weight is added by the returns of the factory inspectors; some private surveys were also undertaken into the use of power in industry. Using the figures is not always straightforward because there were slight changes in the ways in which they were recorded, making comparisons difficult. In particular, the inspectors followed common practice in steam engineering in using nominal horsepower as the unit of measurement until 1850, and indicated horsepower thereafter. Nominal horsepower effectively under-recorded the contribution of steam, to the extent that many engines were more efficient than their nominal rating. At the same time there was some under-recording of waterwheels still working. There were also two principal ways of recording the use of power: one was the number of waterwheels and steam engines, the other was the amount of power that could be generated. The trend towards steam is clear whichever way the returns are reviewed, but there was considerable regional variation.

In the factory inspectors' return of 1838 both

Gibson Mill was built far up the small valley above Hebden Bridge, and had two reservoirs to store water. It proved to be a bit too remote and ceased spinning worsted in the 1890s, becoming a dance hall for half a century. It is now owned by the National Trust, which has developed its water power to make it a flagship site for sustainable energy.

records are included. The textile industry of England and Wales in that year employed 2,230 waterwheels and 3,053 steam engines. In terms of power output, the steam engines were rated at 74,084hp, and the waterwheels at 27,989hp. Big differences in the results from the two sets of measurements are apparent, although waterwheels were in the minority in both. Whereas they represented 42 per cent of the number of engines in use, the amount of power they produced was only 27.4 per cent of the total. Steam accounted for nearly three-quarters of the power output.

Of the two main branches of the textile industry, the cotton industry was rapidly deserting water power. The industry in Lancashire was deriving only 10.7 per cent of its power from water in 1838, but local experience continued to vary from place to place. In the Irwell and Mersey valleys, where coal could be delivered easily and cheaply, water power was already negligible. In the higher Pennine areas north of Manchester there was still a more significant contribution from water. This is demonstrated by the Bolton and Bury districts, where a number of independent surveys complement the returns from the factory inspectors. One of the most detailed was conducted by the mill owner Henry Ashworth for the Manchester Statistical Society in 1837, covering the 'whole district' of Bolton, including the dozen or so parishes surrounding the town. He found nineteen waterwheels generating 511hp, and 90 steam engines producing 2,464hp in the cotton industry. Water power's share was 17 per cent, a small proportion, but well above average for the cotton industry as a whole. The factory inspectors' return confirmed the continuing local importance of water power. For the strictly defined Bolton parish, 24.7 per cent of the power used by the cotton industry came from waterwheels.[272]

The pace of change quickened in the second half of the nineteenth century. The factory returns of 1870 show that only 2.7 per cent of energy consumed in the cotton industry came from water power: no more than 8,390hp were generated by water power, compared with 300,480hp from steam. These returns need to be read with care, however, for there was considerable under-recording, especially of water power. Even with that proviso, the dominance of steam was overwhelming and water power was in retreat by the end of the nineteenth century. The cotton masters of the late nineteenth century were working on a scale much greater than Samuel Greg had done. New mills built in the 1890s were working 90,000 spindles or more, a scale of operation that necessitated steam and, before long, electricity.[273]

It was a similar story in the woollen and worsted industries. The West Riding saw the most rapid growth in the use of steam power of any part of the country. Expansion of the industry as a whole meant that steam power was becoming a major force by the 1830s, with the number of steam engines growing rapidly. There were 316 steam engines in the Yorkshire woollen industry in 1835. Only three years later, in 1838, another fifty-seven steam engines had been installed. Not only were the numbers growing, but also the power of the engine, with the strongest growth being seen in engines of higher power capacity (more than 30hp). It was this that was making steam dominant, for the number of waterwheels at work was changing hardly at all as existing users kept them going. There were 241 waterwheels in the region in 1835; only three of them stopped working in the three years to 1838. This continued use of established sites meant that the amount of water power generated in the woollen and worsted industries was changing only very slowly; indeed, in some respects, the amount of power available continued to increase until the 1860s, as some mill owners invested in new installations and upgraded existing wheels and equipment. The new wheel at Rishworth Mills, Sowerby Bridge, built in 1864, represented one of the last major investments in water power by the Yorkshire woollen industry, but improvement work did continue on a smaller scale for some time afterwards.[274]

As a proportion of the industry's consumption, however, water power was rapidly giving way to steam. The factory inspectors' returns for 1835 record 30 per cent of the power in the Yorkshire woollen mills as being generated by water, and 70 per cent by steam. Elsewhere in the country the pace of change was much less rapid than in Yorkshire. In 1850, water still provided most of the power in the woollen mills of large parts of the West Country, Cumberland, Westmorland and Wales. The returns for England and Wales in 1850 reckoned that 35 per cent of the nominal horsepower used in woollen mills still came from water. Even in Yorkshire there were some districts where water was still the most important source of power – Wharfedale and upper Nidderdale, for example. The introduction of the turbine gave a new lease of life to some watermills, extending their working life into the twentieth century. In Scotland, the woollen industry's adherence to water power was stronger. With good supplies of water, limited competition for water-power sites, and many districts distant from coal supplies, dependence on water power remained strong. In Hawick in the 1830s, only one out of ten mills used steam, and in 1850, 65 per cent of the nominal horsepower in the Scottish woollen industry was produced by water power.[275]

The transfer to steam power was gathering pace, however. By 1871, the last year the factory inspectors made returns of power usage, only 11.6 per cent of the power consumption of the English woollen industry came from water. The proportion in Yorkshire was down to 7.7 per cent, but there was now a general diminution in the use of water power. The woollen industry itself was declining in some of the former strongholds of water power, such as the West Country. Those mills still working in Gloucestershire, however, often continued to use water power, which accounted for 36.6 per cent of power consumption. Other local pockets of water power remained: the blanket makers of Witney were

Tending the spinning machines in Richmond Mill, Huntly, Aberdeenshire, still water-powered in the mid-twentieth century. (James Woodward-Nutt)

Upper Greenland Mill was the last working woollen mill in Bradford-upon-Avon, although it was out of use when seen in 1980.

still dependent on water to a greater degree, as were the mills of Cumberland. The woollen mills of western and northern Wales were most likely to have water as the sole source of power. Their adherence to water power in 1871, at 78.9 per cent, was the greatest, although that was a change from the almost total dependence of 1850. In Scotland the decline of water power had been rapid, from 65 per cent in 1850 to 26.6 per cent in 1871. In northern Ireland, once the change to steam power came, it was almost as fast. The factory returns for 1862 recorded 100 per cent of the power for the woollen industry as still provided by water. By 1871 the proportion had already fallen to 55.7 per cent.[276]

The worsted industry adopted steam power more quickly than the woollen producers. In 1835, water power already accounted for only 21.5 per cent of power consumption in Yorkshire. By 1850 only 13 per cent of the horsepower for the whole of England and Wales was produced by water and by 1871 water power had all but disappeared from the English industry. The Scottish industry kept using water power for longer, with 42 per cent of the power for worsted mills still being derived from water in 1850, but decline was rapid here also, so that by 1871 only 16 per cent of the total horsepower was now derived from water.[277]

Woollen mills used steam for various drying and sweating processes, even for keeping the workers warm, and this hastened the introduction of steam power in those areas most associated with water power. If a boiler was being installed it might as well provide steam for an engine as well as heat. The mill owners of the Yorkshire Pennines were often reluctant to give up using their waterwheels, but by the 1860s those who retained water power were almost all using it alongside steam. Though not necessarily typical in the precise arrangement of usage at the time, Thomas Laycock's mill at Aireworth near Keighley provides one example: it was using water power for five months of the year and steam for the rest.[278]

As the use of steam power advanced in the textile industries, so did concentration in the main centres of Lancashire and Yorkshire. The more rural water-powered industries were unable to expand sufficiently to compete with the larger steam mills. In Wales, the West Country and north-west England, water-powered mills ceased producing woollens and worsteds. The textile industry of Cumbria, producing a range of woollens, cottons and linen, was based around Carlisle, Wigton and Kendal. The decline began as early as the 1840s, as some of the water-powered mills found the competition too great. Even some of the mills further up the Pennine valleys succumbed, being a little too remote. Lord Holme (or Gibson) Mill, a few miles outside Hebden Bridge, ceased production as a textile mill in the 1890s, and became a dance hall and entertainment centre instead.[279]

Paradoxically, the water power that limited the prospects for many woollen mills actually kept others going. The Welsh industry by the 1920s was a shadow of its former self, but there were still more than 100 small textile mills, with low-cost water power retaining its importance. A survey of the rural industries of Wales conducted after the First World War found as many as 62 of the 107 rural mills of the counties of Cardigan, Carmarthen and Pembroke continuing to use water as their sole source of power.[280]

Water Power and the Competition from Steam

In Ireland the supply of coal was, if anything, a greater problem, as it had to be brought across from mainland Britain. Its consequent high price almost certainly prolonged the use of water power, with many businesses using it until mains electricity became an option. Steam engines were being bought in Ireland during the first third of the nineteenth century – 151 were recorded in 1838, half of them in Belfast and Dublin – with many using the local fuel of peat rather than coal. Nevertheless, steam power was barely penetrating the industry of the island at this stage. The availability of water for power remained an important consideration for the siting of new textile mills well into the nineteenth century. It even attracted some entrepreneurs to move across the Irish Sea, including Thomas Crosthwaite, who set up a flax mill in 1810. The woollen industry of Ulster relied on water power at least until the second half of the century. Many of its mills were converted from corn to woollen manufacture. After 1850 steam began to make inroads into the Irish woollen industry; although the progress of steam thereafter was rapid, the adoption of turbines enabled many mills to continue deriving at least some of their needs from water into the twentieth century. Among the late users of water were the Ballygarvey Mill of Raceview Woollen Mills, which found two-thirds of its power from turbines until 1957, and a mill at Derryvale, which was still operating by water power in 1962.[281]

While the main branches of the textile industry were turning to steam, some smaller branches continued to use water power to some extent. The bleaching and dyeing trades, for example, were still significant users of the waterwheel before 1850. Some woollen mills were taken over by other textile businesses, for which the water power was going to be adequate. Silk, hosiery and lace were all users of some of the old woollen mills of the West Country. There were a dozen silk mills at work in Gloucestershire in the mid-nineteenth century, for example. In Dorset and Somerset woollen mills were put to new uses, making sacking, rope and twine.[282]

INDUSTRIES OTHER THAN TEXTILES

One of the first industries to turn to steam power was iron-making. Henry Cort was producing material for the Navy at a water-powered ironworks at Funtley, near Fareham. In 1784 he patented his puddling and rolling process, which enabled wrought iron to be made more cheaply without the lengthy hammering at the forge. He soon discovered, however, that his undershot waterwheel offered too

A Welsh woollen mill photographed in 1937, with a breast-shot wheel fed by a launder. (Museum of English Rural Life)

Water Power and the Competition from Steam

Witchampton paper mill in Dorset was one of those that continued using water power late on. The headrace feeding into the wheel-house was photographed by Alan Stoyel in 1989. (Alan Stoyel)

little power for his furnaces. By the end of the eighteenth century, the iron industry was turning to coke, coal and steam power, in consequence leaving most of its former sites in southern England. The water power was not necessarily abandoned, as iron-working sites were converted to other uses. Coxe's Mill, near Weybridge, for example, was an ironworks until about 1829, when it was converted to a silk mill. It was changed again into a water-powered flour mill, and it remained that way until rebuilding with steam power took place in about 1900.[283]

The paper industry began to move to steam power at the beginning of the nineteenth century. Probably the first paper mills to have steam were the Thames Bank Mills and a mill in Chester, both in 1802. The following year, the Tyne Steam Engine Mill in County Durham opened, and the first steam engine was installed at a paper mill in Scotland, a Boulton & Watt engine at Devanha Mill, Aberdeen. At first, steam was only a direct replacement for water power in driving the pulping machines, and this, initially at least, slowed down the rate at which steam was taken up. The issues of the reliability and low running costs of water power were considered in the paper industry as they were in others. Compared with the capital required for the machinery, the cost of water power was a small part of the total. Steam power also represented a small fraction of the costs, often smaller than water power. With most paper mills being located on a very good site for water power, there was little incentive to abandon it immediately, and investment in water power continued for some time in the early nineteenth century: John Dickinson, for example, installed a new wheel at his Croxley mill in 1826. A good deal of water power was still used in the industry in the 1840s–50s – Shotley Bridge Mills in Durham and the Albury Mills in Surrey, both significant makers, were still reliant on water at that time. However, in order to meet expanding demand and the greater power requirements of newer machinery, steam soon became the answer.

Increased mechanization of paper manufacture in the early nineteenth century prompted the more rapid introduction of steam power, especially after about 1830. The Fourdrinier paper-making machine was developed in the first decade of the nineteenth century, and John Dickinson patented his drying machine in 1817. With the introduction of machine processes and the consequent need for greater power in the mills, the adoption of steam power became more general and more rapid. Dickinson turned to steam in 1815, and by the 1830s most of the large paper-making businesses had followed suit. These firms regarded steam as more reliable for their scale of operation. By 1871, steam provided four times as much of the power used in the United Kingdom paper industry as water did. Even so, paper-making remained one of the major users of water power in British industry, with more than 8,000hp coming from that source. Smaller mills and those producing high-quality hand-made papers often continued with

Water Power and the Competition from Steam

water power for much longer. Taverham Mills in Norfolk, famous in the nineteenth century for making the paper for *The Times*, still had three waterwheels drawing power from the River Wensum alongside eleven steam engines in the 1890s. Bleachfield Mill, Ayton, in Berwickshire used its breast-shot wheel until 1942, while Woodside Mill, Aberdeen, had its wheel, built in 1826 and rated at 300hp, at work until 1962. Turbines often replaced the wheel: indeed, cheap turbine power was held to be the means by which the paper mills at Exeter were enabled to keep going in the 1890s.[284]

Mechanization and steam power prompted the industry to gravitate further towards some of those areas of concentration in Hertfordshire, Kent and Lancashire, while water-powered mills elsewhere closed. The six paper mills in Lincolnshire disappeared during the mid-nineteenth century. The industry of Ireland shrank; it continued to rely on water power more than the factories in the main producing areas did. Even in such established areas as Hampshire, mills at Andover were closing down by the end of the nineteenth century.

For many small-scale industries, water power continued to appeal, even where cheap coal was readily available. The running costs of water could be as low, and this was still a major consideration in the choice of power. Sheffield was one place where this was true, where the many small-scale industries using water power found its low running costs to be of considerable benefit. In the 1860s there were thirty-two water-powered grinding mills in the cutlery trade. A serious flood resulting from weaknesses in a dam destroyed most of the wheels in the Loxley valley on 11 March 1864, but it was still considered worthwhile rebuilding them. Most were restored shortly afterwards and continued to work beyond 1900. The needle-makers of the Redditch district were mainly using water power in 1870, and continued to do so into the twentieth century. At the beginning of the twentieth century water-powered makers of edge-tools in the West Country were still able to compete with larger factory-based enterprises in the Midlands.[285]

The advantage of almost costless power was so great that a number of small businesses were started in the late nineteenth century using mills made available by changes in other industries. Makers of walking sticks, umbrellas and pins took over former woollen mills in south-west England, using them for some decades. Before that, James Coate had taken over a corn mill near Chard, Somerset, in 1847 to make bone-backed toothbrushes and hairbrushes.[286]

The manufacture of gunpowder was another

Churchill Forge, Worcestershire, is on a site used for water power since the fourteenth century. From the eighteenth century at least it was a spade mill, specializing later in ladles for the Stourbridge glass industry, as well as making spades, forks, rakes and other tools. It was working until the 1970s, since when it has been preserved. It has two waterwheels, one of which is being inspected in this view from the 1940s. (Museum of English Rural Life)

Water Power and the Competition from Steam

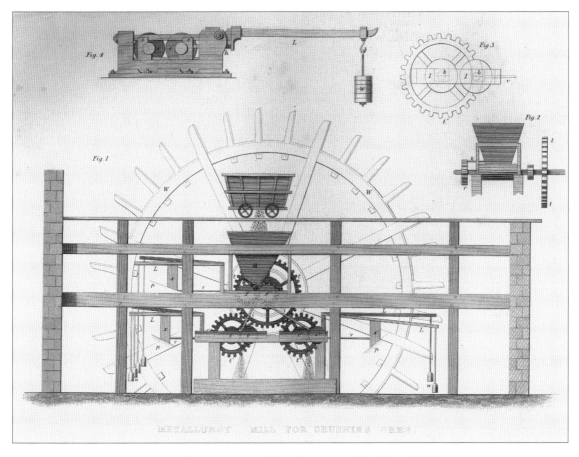

A water-powered ore crusher, an illustration from Tomlinson's Cyclopaedia. *(Museum of English Rural Life)*

industry in which water power remained important. The factory return for 1870 quoted figures of 615hp for water and 769hp for steam power used in the industry in England, but at least eight works were not included. Most of them were water-powered and their inclusion might have doubled the figure for water.[287]

Mining still had use for water power for pumping. In the Cornish mines there were about 650 steam engines producing about 40,000hp in the 1860s, but also 200 waterwheels still at work, producing about 5,000hp. Devon similarly had several waterwheels at work in the final decades of the nineteenth century, thirty-three of them employed by Devon Great Consols.[288]

FLOUR-MILLING

Steam, meanwhile, had been introduced to flour-milling too. In 1788 the Albion Mills in London were opened. Boulton & Watt had a 20 per cent share in the business, so naturally it was one of their new steam engines that provided power for the twenty pairs of stones, which were able to produce ten bushels of flour an hour. This, the most high-profile of a new generation of steam mills, seemed set to revolutionize corn-milling. It was a bright prospect for its promoters, but to the traditional millers along the rivers near London it was a cause for concern, even fear. Arson might have been the cause when the Albion Mill burned down in 1791, and suspicion fell upon some of the watermillers who had complained about the newcomer.[289] The fire certainly removed a

The wheel built for the Weardale Lead Co., at Kilhope, Co. Durham, where it was used for washing ore. This is how it looked in derelict and vandalized state in 1962. It is now preserved. (Museum of English Rural Life)

competitor, but in fact the mill had other technical problems, which meant that the watermillers for the time being had less to fear than they thought.

The Albion was not the only one of its type: between 1780 and 1800, forty-five steam mills are known to have been built. They were in places that were very short of water power, such as the west Midlands and especially Birmingham, places with good supplies of coal, such as the north-east, and places where coal could readily be brought in, such as Southampton and Oxford. The potential of these mills was enormous. One at Dartford could grind 300–400 quarters of grain a week, about as much as a watermill would grind in a year.[290]

Productive though they could be, there was still need for further development before steam power for milling became established. The steam engines of the late eighteenth century were not always suited to providing power for grain-milling, which was one of the problems faced by the Albion Mills. This first generation did not presage an immediate rush of investment in steam mills that would quickly put watermills and windmills out of business. Far from it: many of the first steam mills were short-lived. Mills at Oxford, Southampton and Liverpool closed after no more than two years. A mill at Chesterfield was advertised for sale in 1798, described as having been 'formerly used for corn, now for grinding cast metal forks'.[291] No wonder the owner of the Mersey Mills at Warrington could confidently claim in 1814 that London would always need water power: 'Perhaps no place in the world is worse situated for grinding flour by steam.'[292] In the longer term he was wrong, but for the time being water, and wind, could be competitive with steam power. It was not until the railways were able to transport coal more cheaply to the inland towns, where many steam mills would be built, that steam began to provide an effective competition for water and wind power. This was especially true of southern England, furthest from the coalfields: few steam mills were built here before the 1840s.

Steam mills had a better chance where water power was in short supply. Ten mills had been built in Birmingham by 1827, with a capacity to supply more than half of the local population. By this time, with engines more suitable for milling available, steam power was starting to make an impression in some of the bigger towns, especially in the north of England where coal was available cheaply. It was in such places as Newcastle upon Tyne that most of the early steam mills were built. In 1822 there were eight steam mills in the city, none of them large, but together they had the capacity to mill 880 quarters of wheat each week, almost four times the output of the local watermills. Forty years later, in 1863, the impact of steam was marked: two large steam mills now accounted for two-thirds of Newcastle's flour production.[293] Where coal was not so cheap, steam mills still struggled. In

Hereford and in East Anglia there were mills built in the 1820s that lasted a very short time.

As the railway network expanded, carrying coal further afield, new steam mills were being built in the mid-nineteenth century at places with good rail connections, often the market towns, such as Chelmsford. The business of E. Marriage & Son had been built up at East Mill, Colchester, one of the big watermills around this town, but further expansion was achieved with the adoption of steam and the rebuilding of the mill to accommodate the engines and roller machinery from the mid-nineteenth century onwards.[294]

However, the growth of milling under steam was not yet developing at a rate fast enough to have much effect on the numbers of water corn mills. Steam was providing additional capacity more than acting as a replacement. In the 1860s and 1870s most watermills, to all appearances at least, were as busy as ever, as countless local studies indicate. The output of the watermills of Hampshire in 1880 was only a little less than it had been in 1850, and a little greater than it had been in 1800.[295] Quite a few watermills had added a steam engine alongside the existing mill, a common course of action taken to give greater operational flexibility but a sign also that there was sufficient concern about the new steam mills to justify the investment. At the same time there was continuing investment in new waterwheels and gearing, often of iron, an indication that there was still reasonable confidence in the prospects for watermills.

In general, the effect of expanding demand was that in some southern counties of England, Dorset for example, there were probably more mills grinding corn in 1870 than there had been a century before. In that year, two-thirds of the power available for grain-milling in England and Wales was still provided by water power. In northern Ireland, where the high cost of imported coal confined steam mills almost entirely to Belfast, the proportion was even higher. Even though the number of mills in Ireland had been set on a downward course following the famine, the effect on the total was still modest by 1870. As late as 1915, water power was used by 85 per cent of the corn mills of Ulster. When it came to total output, it was a different matter: a few steam mills could more than outweigh the production of watermills.[296]

What really changed the fortunes of the water corn mill was a change in the supply of wheat. By the end of the nineteenth century more than half the British consumption of wheat was imported. The cost of transporting grain in bulk from large-scale producers in America and elsewhere fell during the 1870s, and with it the price of cereals came down dramatically. To process the supplies of wheat arriving by the ship-load, new mills were built, and these were large ones powered by steam and later by electricity. They were sited mainly at the ports, such as Southampton, Hull and Bristol, so that grain could be easily unloaded direct from ship to mill. Large mills required large sums of capital and flour-milling was becoming a big business dominated by new expanding firms, such as Spillers and Ranks.

The new mills at the ports used the developing technology of roller-milling. Grinding grain between porcelain rollers instead of stones had been developed during the 1860s, first in Hungary, then the United States. Roller-milling had greater capacity than stone-grinding; it was more efficient, using less power per unit of output, and it allowed much greater refinement in the process of milling. As roller-milling technology was taken up in Britain, during the 1870s, steam mills at inland towns were among the early users, but, coinciding as it did with the rise in imports of grain, it was the obvious choice for the new mills at the ports.

Roller mills could work satisfactorily under water power, and a number of millers installed them as a means of countering the competition. The water-powered Abbey Mill at Reading was one of the first in Britain to install roller machinery. Although it was driven by steam, the waterwheel was retained to power three sets of stones that were kept to grind barley. Other sites, however, did use water power with their roller mill. Town Mills, Newbury, continued to use water power after roller plant was installed in the 1890s and Edington Mill at Hungerford did the same after converting to roller machinery in 1898.[297] Osney Mills at Oxford also installed roller mills in the 1870s.

Water Power and the Competition from Steam

Caudwell's Mill, Rowsley, Derbyshire, one of those that installed turbine-powered roller mills.

Often, turbines were preferred to waterwheels. Wear Gifford Mill, on the River Torridge in north Devon, was given a thorough modernization, with a 30hp turbine providing power for Simon roller mills, all the ancillary machinery and for generating electricity to light the building. It proved enough, with the turbine renewed in 1910, to keep the mill working until 1940. Caudwell's Mill at Rowsley in Derbyshire installed rollers in 1906, driven by a turbine wheel, which it continues to use. Some manufacturers made roller mills, such as the 'Midget', designed with the small miller in mind. This model could produce 11 stones of flour per hour driven by a waterwheel, and was installed by a number of watermills.[298]

It was a small minority of watermills, however, that was able to gain from the installation of turbines and roller mills. The large mills driven by steam or electricity, which could offer greater control, continuous power and almost unlimited expansion in the scale of milling, still had the competitive edge. The economies of scale enjoyed by the port mills meant that they could supply flour at retail prices far lower than that offered by the rural mills. Since most of them were inland, the country mills had limited access to imported grain. At the same time local supplies were decreasing as cereal growing in Britain declined in the face of the imports; between the 1870s and 1890s, the sowing of wheat fell by approximately half.

The effect of all these developments – the large port mills, the wheat imports and the introduction of roller-milling – was a rapid fall in the number of small mills. Joseph Rank started out in business as a wind-miller at just the wrong time, in the 1870s, when there was already overcapacity in milling around Hull where he lived. Rank had the ambition and entrepreneurship to move on to become an owner of port mills, but most millers relying solely on wind and water were forced out of business. In 1880 there were about 10,000 rural mills, using water and wind, in Britain. By 1910 the number had fallen to about 2,000. Local experience confirms the national trend: from 190 watermills in Hampshire in 1880, the total had fallen to 116 by 1900, and again to seventy-six in 1914. Decline proceeded more slowly in Ireland, with 1,200 mills remaining in 1900 compared with the 1,800 of 1870. The power capacity of the watermills had almost halved, however, from an estimated 60,000hp in the 1870s to 38,865hp (out of a total for the flour-milling industry of 177,451hp) recorded in the census of production of 1907.[299]

The journals of the trade, *The Miller* and *Milling*,

Water Power and the Competition from Steam

A. R. Tattersall & Co., of London, sold the 'Midget' roller mill, which they were confident would have dramatic effect on the fortunes of its users. This is from a catalogue of 1915. (Museum of English Rural Life)

reflected the changes. In the 1880s they still contained some discussion of the best types of water-wheel and turbine. By 1900 watermills are barely mentioned, except in the advertisements of mills for sale, which were as likely as not to be sold for the quality of the house and the fishing as for the mill.

The surviving watermills kept going by meeting local demand, sometimes for flour, but mainly for animal feed. Feed-milling became a lifeline for the water-, and wind-, miller. Grinding grain for

Water Power and the Competition from Steam

Bryncrug corn mill, Merionethshire, built in the sixteenth century, was still in use in 1938, run by Miss Lois Owen. (Museum of English Rural Life)

Mr Lightfoot, the last miller at Hessenford, Cornwall, in 1957, when he retired aged 75. (Museum of English Rural Life)

livestock feed had for a long time been a part of the country miller's business. Some were specialists at it, and in some places, Northern Ireland for one, there was a distinction made between the flour mill grinding for human consumption and the grist mill grinding feed. Most mills in England would grind both: a mill with three pairs of stones would typically use two for flour and the third for feed. By 1901 most watermills had become grist mills. In that year, of 7,000 mills in England and Wales grinding by stones, 5,500 were producing feed only. Demand for their flour had gone, but there was continuing (and in some respects growing) demand for animal feed. Cattle had become a more important part of farming, and farmers would find at least some of their feed from home-grown cereals, providing work for the rural miller. This was what kept most of the working watermills and windmills of the twentieth century going; almost all those that survived the Second World War were grinding oats and barley for livestock food.

It was not plain sailing in the feed business either, for any farmer could buy himself a cheap small grinding mill to grind his own feed, using horse power or a small oil engine. Many of the watermills that kept at work beyond about 1920 did so through sheer doggedness on the part of the miller. When he retired so did his mill, and an abundance of local stories bears testimony to this. But there were some that had remarkable success, at least for a while, having a good local market. They might have installed additional power – steam in the nineteenth century, diesel or electricity in the twentieth – and some roller mills, but more than a handful continued with water and were still active after 1945. For these it was force of circumstance that eventually stopped the mill: retirement of the miller with no successor, flood damage, a repair bill beyond the capacity of

EXTERNAL PRESSURES ON WATER POWER

The watermill at Aymestrey, Herefordshire, being restored for use during the Second World War. (Museum of English Rural Life)

There had always been competition for the use of water. The traditional rivals – other mill owners, farming, fishing and navigation interests – were still present, but industrialization and the growth of towns during the eighteenth and nineteenth centuries introduced new demands for water. Canals were threatening to cut across the rivers supplying mills; there were new industries wanting to extract water in large quantities; above all there were increasing urban demands for water supply and sewage disposal. All represented a threat to the survival of water power, but, while it was commercially active, efforts were usually made to accommodate it. It has been a different matter with the decline of water power in regular use.

Northern towns began to build reservoirs in the Pennine hills to provide their water supply during the eighteenth century. The increased demands, from urban water supply and from the effect of drainage works on rivers, was remarked upon by William Fairbairn in his *Treatise on Mills and Millwork* in 1861 as a serious competitive pressure on water power. Reservoir-building for urban supply gathered pace throughout the nineteenth century, until the number reached nearly 200 in the 1890s.

Extraction of groundwater had been of minimal impact before large-scale urbanization, confined mainly to local wells, but nineteenth-century demands for public water supply and sewage disposal soon put pressure on underground sources. Soon after the passage of the Public Health Act 1848, Croydon established its Local Board of Health, which developed ambitious plans to tap groundwater. Landowners and operators of mills on the River Wandle were alarmed. They challenged the Board's rights to the water in a legal action that became known as 'The Great Water Case' or, more prosaically, *Chasemore v. Richards*. It proceeded up to the House of Lords, where judgment in 1859 went against the miller in whose name the case was brought, with the decision that no one had absolute right to water below ground. The case had some significance for river management, for it opened the

the business, or impositions of water boards.

There were some exceptions to the rule. Some places were sufficiently remote for a local monopoly to arise, and this helped watermills to keep business, even for flour. The Meon Valley, Hampshire, was remote enough for two watermills, at Wickham and Droxford, to be kept busy still grinding wheat at least until 1914. Huntley & Palmer's biscuit factory at Reading used large quantities of locally grown wheat, and this specialist demand kept a number of mills on the Thames, Kennet and Loddon rivers at work. Some of them modernized with steam power and roller mills, but many kept their water power.[300]

War-time demands brought some watermills back into use, usually for feed-milling, during the 1940s. This happened in Northern Ireland, for example. Few survived for long once peace returned and it did not take much for the mill to close for good. The freezing winter of 1947, for example, stopped a number of waterwheels turning. Lealholm Mill, in the North Yorkshire Moors, was one that did not restart, its trade having drifted away.[301]

way for authorities to draw from underground sources with little impediment. It also represented a shift in public attitude, placing urban demands above those of users of water power, and this was to become more pronounced during the twentieth century. At the same time in the mid-nineteenth century, urban expansion was beginning to cause concern about the quality of water available to mills. Mill owners on the Wandle, again in the vanguard in experiencing the effects of increased urbanization, were expressing anxiety about water polluted by the effluent of Croydon.[302]

As the mains water supply was extended, its needs were often accorded priority over those of mills. Reservoirs built to store water for the public supply and groundwater extraction from rivers could seriously diminish the flow to mills downstream. Water authorities all over the country engaged in such works. Froghall Mill on the River Churnet in Staffordshire had much of its water supply drawn off for municipal use, and then a new reservoir built by Staffordshire Water Board further restricted supply. Despite this, the mill kept working until the early 1960s. Eastbourne Water Works made boreholes in the valley of the Cuckmere river that drained away most of Hellingly Mill's water supply.[303]

The Water Resources Act 1963 had public water supply as its priority. The twenty-nine regional river authorities for England and Wales established by the Act were to 'manage and initiate water conservation', and as part of that they were to regulate the consumption of water. The principle of licences and charges for abstracting water from the river was introduced, including 'temporary abstraction'. Users of water power were incensed: they were being charged for something that had hitherto been free, and for 'consuming' water when they were returning it to the river after it had been through the mill. The government argued that there would be no detrimental effect on the users of water power, but experience did not necessarily bear that out. By the mid-1970s, these users were raising serious concerns, with claims that the Act had all but killed off the use of watermills in some places, such as Cumbria. Under pressure from a new organization, the National Association of Water Power Users, a review of the system was instigated, which resulted in its effects being mitigated. The basic regime of licences and regulation of consumption has remained, however. After a number of organizational changes, management is now in the hands of the Environment Agency.[304]

Weirs and millstreams had not always been popular with farming interests who wanted free drainage. As working mills declined in number, river authorities began to support that interest by destroying dams to ensure that mills could not work again. In 1937 the Avon & Stour Catchment Board ordered the removal of the hatches on the river and floats on the wheel at Wimborne sawmill, in order to maintain free drainage. The weir serving the small Felin Lyn Mill near Chirk had suffered some damage, and the river authority in Denbighshire had it completely destroyed because it was considered an obstruction to flows for water and fish, thus ensuring that the mill could not be brought back to work. The existence of a mill in working order was no guarantee of protection. Welsh Water in 1983 lowered the weirs serving Priory Mill, Usk, 'as part of a £1 million scheme to prevent flooding in the area'. The mill was still used occasionally to give demonstrations to schools and others, but, because it was not a commercial operation, the water authority overrode its claim to supply. Improvements to field drainage could also restrict the supply of water to mills. Instead of seeping through the ground, eventually reaching springs, and thence the steady flow down the stream, there was a rapid run-off from fields to rivers. Another effect of drainage improvements could be that the flow of water increased sufficiently when mill sluices were opened to create floods in neighbouring fields, something not encountered before. A number of mills were forced to stop work because of this. The wheel at City Farm, Sydling St Nicholas in Dorset, had to stop work in 1915 because lowering the hatch to run it caused flooding in nearby houses. Another Dorset farm, Preston Farm, at Tarrant Hinton, abandoned the use of water power in 1940 when the flooding of meadows became too much.[305]

12 Landscapes of Water Power

According to one twenty-first-century geographer, 'There are few entirely natural rivers left in Britain.'[306] One of the main reasons for that is the effect that water power has had on the landscapes of rivers and streams. Transport, agriculture, fishing and water supply have all played their part, but the many thousands of millwheels have had an enormous impact. On most rivers, development and redevelopment from medieval times onwards have created a palimpsest of waterworks.

Even though most of the mills, mines and forges that used the power have been swept away, left derelict or converted to other uses, their effect on the landscape remains, in the millstreams, millponds, dams and weirs, in some of the buildings and sometimes in whole settlements.

Much of the time, constructions such as mill leats go unnoticed: so long is the history of water power that they can appear to be natural features. Some, like the Costa Beck in north Yorkshire, have acquired a name as though they were natural streams. This 'beck' is actually the diversion from the Oxford Beck, which was created to feed the mills of Costa High and Costa Low. The 5-mile diversion from the River Kennet, cut to serve Reading Abbey Mill, is known as the Holy Brook; flowing through the meadows beside the motorway, it looks for all the world as though it had been created naturally.

Throughout the land, the impounding of water and the artificial cuts have affected the natural flow of streams, by the way they erode and deposit sediment. Even in decline water power has its effects. The wheels have stopped turning, and so the regular maintenance of watercourses and weirs has ceased. A millpond in use is regularly dredged, but when this no longer occurs it accumulates sediment, which can make the pond a valuable haven for wildlife. Conversely, the deterioration of neglected dams

The country mill: Ramsbury, Wiltshire, in the 1930s. (Museum of English Rural Life)

Landscapes of Water Power

The Holy Brook flows through Calcot on its way towards Reading. It will shortly reach Calcot Mill, the first of its calls on the way to the Abbey Mill.

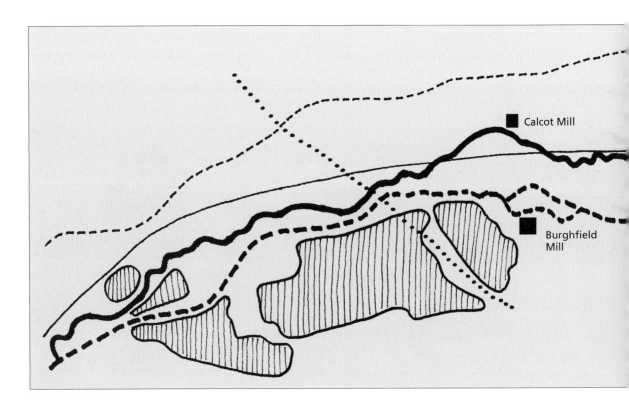

Landscapes of Water Power

could adversely affect the flow of the river. These are factors that have played a part in recent thinking on the management of rivers.[307]

Following the demise of water-power users such as the Wealden iron industry or the lead-mining of the Pennines, it may seem that little remains on the surface. Mine buildings and winding gear have been demolished. In the Weald there are the reservoirs – the hammer ponds – but many of those have been drained. However, the signs of the past are often there. It requires, perhaps, something of the archaeologist's eye, but it is possible to spot, for example, the earthworks that are the remains of the dams that held back the ponds, and the channels, some now dry, others still flowing, that are diversions off the river.[308]

In the countryside the characteristic landscape created by water power is the linear development of

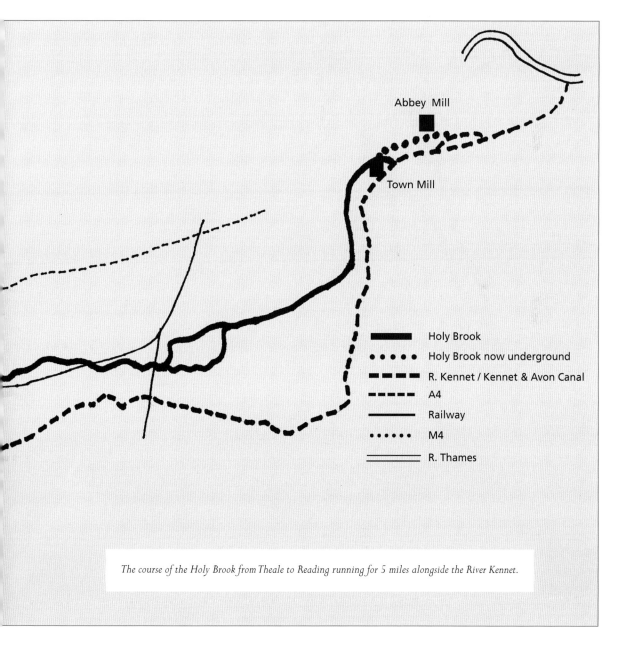

The course of the Holy Brook from Theale to Reading running for 5 miles alongside the River Kennet.

Landscapes of Water Power

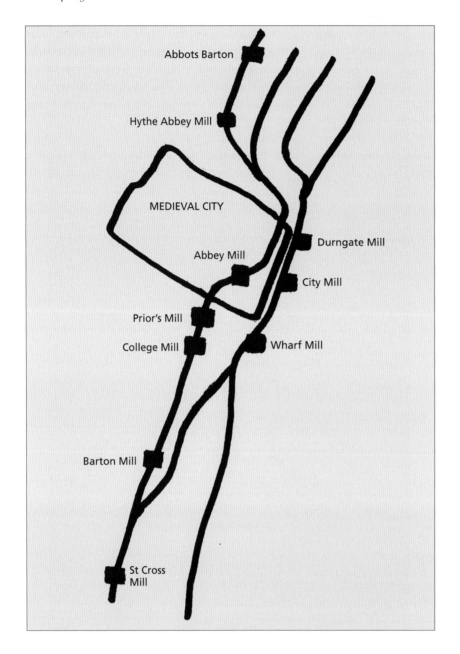

Map of the River Itchen and its branches and the siting of mills around Winchester, showing the extent of the medieval city. Not all the mills were working at the same time: some medieval mills were closed or replaced; some mills, such as City and Durngate, had a very long working life.

rivers punctuated at intervals by mills. Weir, leat, mill, mill house, and often a river crossing to enable wagons to reach the mill, followed one after the other. On middle and lower stretches of rivers, where the flow was more gentle, more needed to be done to maximize the power. Channels serving mills tended to become longer, until there were some that were bypassed for more than half their length. This could result in such extensive works as the long bypass channel to serve half a dozen mills built by George Robinson in Nottinghamshire.

Water power was no less influential on the landscape of towns. Where several mills were squeezed into the crowded urban space, the watercourses multiplied, and they were confined within strictly defined artificial channels, often taken underground. Development in Winchester began in the medieval period. The course of the Itchen was diverted both above and below the city to serve ten mills between Abbots Barton and St Cross. One channel was taken through the walled city to feed the abbey mill, while the main river went just to the east of the city wall, with leats taken off for Durngate, City and Wharf mills. Most of the mills have gone, but the multiple channels of the river still flow.[309] Similar development occurred at Exeter, where a network of leats crossed Exe Island to serve a number of corn mills and fulling mills, around which new settlement developed outside the city walls.[310]

Growth in the use of water power since the medieval period added to the pressures of urban development. 'A little over a century ago Wandsworth was little more than a settlement of watermills on the Wandle,' R. Thurston Hopkins wrote in the early twentieth century. The wheels turned for a wide variety of processes: iron forges, paper mills, white lead mills, dye works, calico printing and bleaching. Before the expansion of London's suburbs enveloped the area, the River Wandle was said to have 200 waterwheels at work. Among them, one served the textile-printing works of the London store Liberty at Merton. This was next to a calico printer's bought by William Morris & Co in 1881. Carshalton was a local centre, with about ten mills at work between 1800 and 1850. Four of them were flour mills, all of reasonable size with at least three pairs of stones. Besides these were two leather-dressing mills and works for snuff, paper and hemp-spinning. Of the neighbouring Tillingbourne river, John Evelyn, a local landowner, wrote in 1675, 'I do not remember to have seen such a variety of mills and works upon so narrow a brook and in so narrow a compass.' In little more than 10 miles of the brook's course eastward from Shalford there were at least twenty-two mills, producing, at one time or another, gunpowder, paper and iron, as well as corn. Uxbridge supported the greatest concentration of mills in Middlesex, mainly flour mills, whose trade expanded in the late eighteenth century after the opening of the Grand Junction Canal.[311]

The growth of industry in the eighteenth century created new concentrations of water-powered mills and led to competition for sites along the streams of the Pennines and Peak District. The use of water power for cotton mills took industry to new places, most of them rural, and the greater scale of factory-based production introduced large new buildings into the countryside. With steam power the trend swung in the opposite direction, taking mills back into the larger towns, but many water-powered mills on the best sites continued in operation throughout the nineteenth century, and several of them remain in some of the Pennine valleys, for example.[312]

Quarry Bank Mill in Cheshire was a prime example of the large new mill in a sylvan setting, although today it is just a stone's throw from Manchester airport. This was where Samuel Greg established his cotton-spinning enterprise in 1782. He chose a site near a spot where a water corn mill had been established in 1362. At this point the River Bollin, shortly after being joined by its tributary the Dean, passes through a gorge. It offered the miller very usable water, but supply was not guaranteed,

A row of cottages built near Quarry Bank Mill to house workers. (Museum of English Rural Life)

Landscapes of Water Power

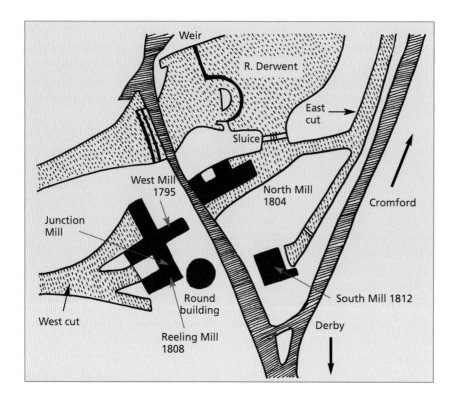

Something of the impact of water power on the river landscape at Belper is revealed in a sketch map of the mills and their weirs and water channels.

with droughts being experienced sometimes in spring and early summer. The new mill Greg built was extended several times until it accommodated more than 10,000 spindles.

As the mill grew so did the need for workers, so that water power became responsible for the creation not only of the factory but also of the surrounding settlement. Samuel Greg set out to create a community for his workers. He built a house for the apprentices in 1790, and cottages for adult workers on the site near to the mill. The development of the village of Styal followed, but it remained relatively isolated and rural, despite being only a dozen or so miles from Manchester. Samuel Greg junior followed with a settlement to house the workforce for another of the family's mills in Cheshire, Lowerhouse at Bollington, for which he had more utopian ideals.

Similar ideals inspired Robert Owen when he built a model community to support the mills at New Lanark. This was another town created by water power, a place identified by Arkwright as suitable for cotton-spinning when he visited in the 1770s. The first mill was built in 1784, followed by several more during the next fifty years. All were built along the River Clyde. A channel ran behind, taking the water from the river to each mill in turn. Water continued to provide power until the closure of the mills in 1986. Three water turbines were then in use: one of 450hp at no. 1 mill, one of 650hp at no. 3 mill, and one of 88hp at Dye House. Full power was achieved only when the river was in full spate.[313]

Existing towns also grew under the impact of industrialization: Congleton, for example, a centre for the water-powered silk industry, increased in population from 4,378 in 1801 to 11,439 in 1831. Richard Arkwright's mills in Derbyshire stimulated new development at Cromford, although it remained a small settlement. By 1881 its population of 1,074 had already fallen from the peak of 1,291 reached in 1831. It had, however, been developed, with new housing and a market introduced by the Arkwrights.

The Strutts' town of Belper was on a far larger scale, with streets of new housing being built to accommodate the workers at the mills. It had a population of 5,778 by 1811, and continued to grow, reaching 8,527 in 1871. This new township was supporting mills on the River Derwent. These were some of the largest water-powered mills in the country and had a significant impact on the local topography.

The first was built in about 1778, and by 1813 that mill had been rebuilt and four more added, creating a complex of water engineering to serve them. The river above the bridge was widened with the construction of the great semi-circular weir, and parts of the channel secured by retaining walls. A network of leats led off to the mill wheels. The weirs and flood-gates, according to John Farey, 'give a more perfect command of this large and very variable river … than can perhaps any where else be witnessed'.[314] The whole development drew admiring descriptions. Andrew Ure in his *Philosophy of Manufactures* in 1835 thought that Belper 'with its river, overhanging woods, and distant range of hills … had the picturesque air of an Italian scene'.[315] Arkwright's Masson Mill was also impressive. The strength of its architecture prompted favourable comment from William Adam in his topographical book *The Gem of the Peak* in 1838, while Erasmus Darwin had been inspired to write verse in praise of this great industrial achievement. But not everyone approved. Sir Uvedale Price in *Essays on the Picturesque* (1810) wrote of Masson, sited in the narrow limestone gorge of Matlock Dale, and of mills like it, that 'nothing can equal them for the purpose of dis-beautifying an enchanting piece of scenery'.[316]

Even in the smaller places, water power had its impact. The large numbers of small mills in such places as the Lake District, each with perhaps no more than a dozen or so workers, hardly constituted major employment. Yet the range of industry attracted to the area in the eighteenth and nineteenth centuries by cheap water power – gunpowder, snuff-milling, saw-milling, paper-making, spade manufacture – together made a considerable difference to the physical development of villages, and their economy and social structure, until at least the last quarter of the nineteenth century.[317]

The importance of water power for local economy and society was perhaps most marked in upland Britain. In the lead-mining districts of the Pennines complex networks of water channels and reservoirs were constructed to feed the waterwheels at the dressing floors and pits, which were prominent features of the nineteenth-century landscape. High up on Grassington Moor, for example, the Priest's Tarn was one of the principal new reservoirs built to store water at the head of the streams; the Duke's Watercourse was an entirely artificial cut linking some of the waterwheels. Priest's Tarn remains, but following the decline of the industry subsequent re-working of the land has left little trace of the industrial landscape.[318]

The impact of water power on the landscape has generally been regarded as benign. The exception has been large-scale hydro-electric power generation. Most of the power stations have been built in the regions of mountain and moorland, which by the twentieth century were almost universally regarded as areas of natural beauty. Defenders of the landscape rose against the intrusion of the power stations. The power stations themselves were said to be ugly, but greater opposition was focused upon the dams and the large reservoirs behind them. The power stations were on a scale far greater than the largest of the Industrial Revolution's water-powered mills, and their demand for water has been proportionately great. To meet it large dams have been built and extensive tracts of land have been flooded. Opposition came from a number of quarters. Sportsmen in the Highlands wanted the wild landscape to be preserved and fishermen were concerned about the effects of dams and power stations on salmon rivers. Others were concerned about the effects of large storage reservoirs on the ecology of the river system, the turning of a once-large stream into a trickle as it flowed away from the dam. Yet others wanted all 'natural' landscapes to be preserved.

The combined efforts of the interested parties succeeded in preventing some Scottish hydro-electric power developments before the Second World War. After the war, the nationalized authorities pushed ahead with development, although opposition had far from diminished. There was intense campaigning against hydro-electric power

Landscapes of Water Power

Modern environmental landscaping at Glendoe reservoir. (Scottish and Southern Electricity plc)

schemes in both Scotland and north Wales in the mid-twentieth century. The publication in 1949 of proposals for large-scale schemes in Snowdonia and mid-Wales by the new British Electricity Authority prompted protests from the Council for the Protection of Rural Wales against the effect of large dams and water pipelines on the landscape. Clough Williams-Ellis, its president, and an architect and author, wrote *Headlong Down the Years* with his wife Amabel, a polemic against the power schemes. Welsh Nationalists complained that much of the power generated would go to England. Of all the developments, the pumped storage scheme at Ffestiniog attracted some of the strongest criticism. Its big new Tan-y-Grisiau reservoir, flooding a deep valley and the path of the Ffestiniog Railway, which enthusiasts planned to restore, created long-lasting embitterment towards the Central Electricity Generating Board.[319]

The scheme for a new power station at Dinorwic in the 1970s prompted another campaign to prevent, or at least modify, its impact on the environment.[320] The opposition by this time was less fierce than before, as much as anything because the power generators had to take their opponents seriously and design power stations to fit the landscape. Many have proved to be less obtrusive than those of the first generation. Many of the concerns of the fishing interests were addressed by the construction of salmon ladders. The opening of visitor centres at such places as Pitlochry, allowing people to see the work of power generation and the salmon passing, resulted in some power stations becoming tourist attractions.

The late twentieth century's green consciousness to some extent has swung the pendulum in conservation circles back more in favour of hydro-electric power. It is a renewable source of energy, and one which, in the opinion of many, is preferable to the more obtrusive wind turbines. This is not a universal view, however, and it seems that, for every green campaigner in favour of hydro power, there is another who still regards it as an invasion of the landscape. Fishing interests remain on the whole opposed, arguing that the introduction of turbines into rivers could adversely affect the migration of species such as trout and salmon.

13 Hydro-Electricity

The generation of electricity from water power seemed to many in the late nineteenth century to be a dream come true. Here was power that was going to be cheap and clean, and not dependent on uncertain foreign supplies. All over the world the utopian attributes of renewable 'green' energy were already being ascribed to it. In Germany hydro-electric power was greeted with enthusiasm as the 'white coal' that was going to usher in a new industrial revolution.[321]

Hydro-electricity came into its own with the high-speed turbine wheel. The vertical waterwheel can do the job, and there were a number of early hydro-electric power projects that used it. Sir William Armstrong installed one at Cragside, and the generator at Mary Tavy in Devon also used a vertical wheel. This last is one of a number of examples in use today, some with old wheels, a few newly built. They operate on the modest scale, however. One typical example was the corn mill at South Wingfield, Derbyshire, which turned to producing electricity for the neighbouring farm using the existing wheel, and was still working in this way throughout the 1980s.[322]

For the generation of electricity on the large scale the turbine is needed. It was not long after the turbine had been developed, in the mid-nineteenth century, that it was first applied to electricity generation. The first instance was in 1869 at Lancey, in France, for industrial use. The first hydro-electricity generating station for public supply was also in France, opened in 1882.

This overshot waterwheel was part of an electricity generation scheme started in 1914 by the Glanllyn estate in Merionethshire. This view is from 1949. (Museum of English Rural Life)

Hydro-Electricity

Hydro-electricity was introduced to the British Isles in 1880 at Sir William Armstrong's house, Cragside in Northumberland. Like many others, Armstrong had been interested in extracting more power from water and at greater efficiency than the vertical waterwheel: 'I thought how great would be the force of even a small quantity of water if its energy were only concentrated in one column.' This line of thinking led him to develop the hydraulic engine to operate cranes and swing bridges, including the one at Newcastle upon Tyne and London's Tower Bridge. In the 1860s he was predicting that electricity would be the power of the future, in succession to coal; it would be generated by water, he said: 'Whenever the time comes for harnessing the power of great waterfalls, the transmission of power by electricity will become a system of great importance.' He put some of his ideas into practice at Cragside. He dammed the Debden Burn to provide a fall of 35 feet to drive a water turbine – a 9hp (6.75KW) turbine built by Williamson Brothers. It generated power for lighting the house, the equipment for which was supplied by Armstrong's friend Sir Joseph Swan. The reservoir of water also operated hydraulic machines in the house, including a passenger lift and a roasting spit.[323] Following this example several private houses and estates in Scotland, Wales and Northern Ireland installed electric lighting with power from water turbines. At least fifty were installed between 1881 and 1894. In the south-west of England, Chargot House, near Luxborough in Somerset, was another of the pioneers, with electric lighting being installed in 1890. It was a long-lived installation too, continuing in use until the thaw after the snow of 1963 undermined the embankment to its reservoir.[324]

Interest in the generation of electric power by water for public distribution was also growing. Between 1881 and 1894 at least fifty-three proposals were made for the construction of hydro-electric power stations in Britain. Only eight of them were put into commission, the first at Godalming in Surrey, in 1881. It might seem an unlikely place, but the same could be said of many of the other schemes of this time. An existing watermill, Westbrook Mill on the River Wey, was converted to produce electricity. It was not a successful enterprise, partly, perhaps, because it used the mill's breast-shot wheel. Instead of installing a turbine, however, a second wheel was put in to provide additional power; in 1884 the project was abandoned. Greenock in Scotland opened a public generating station in 1885 using a 30KW turbine, but that, too, lasted only two years. Further early examples of public supply were at Keswick on the River Grant, started in 1890, and at Worcester on the River Teme in 1894. Uncertain supplies of water meant that both these had auxiliary steam generators. Devon was the English county with most hydro-electric schemes. At Lynmouth in 1890, Charles Geen took power from the East Lyn river to drive a turbine made by Charles Hett, which supplied the town. The popularity of the new electricity supply prompted enlargement of the system in 1895, with a pioneering pumped storage system to use off-peak power to take water up to a new reservoir at the top of the hill above the town. Two Pelton wheels provided additional generating capacity. With auxiliary power from steam and, later, diesel engines, this system continued in use until it was destroyed by the floods of 1952.[325]

Many early hydro-electric schemes were begun in order to provide power for local industries, with surplus power then being offered for public supply. The introduction of the national grid in 1926 stimulated the promotion of more extensive schemes for public supply. The grid provided a more accessible market for hydro stations, which could exploit their ability to switch power into the grid at short notice.

Industrial need was the origin of the first public generating station in Northern Ireland. It was built in 1896 by J. E. Ritter to provide power for his family's mills and remained in private ownership until 1936, when the Electricity Board for Northern Ireland took it over.[326] A similar course was followed at Okehampton in Devon, where a water turbine was installed by Henry Geen, brother of Charles, to work his sawmill. Surplus power was sold to the town, and the whole generating plant was taken over by the West Devon Electric Supply Company in 1930. The company replaced it with a new hydro station at Mary Tavy in 1937. The brewery at Wickwar in Gloucestershire installed a generator worked by an overshot wheel in 1888, and this, too, provided electricity for the town. In Wales some of the quarries were turning to hydro-electricity for their own power needs and building their own water-turbine generators by the early 1890s, while many small-scale installations for industrial or farm use were made in Ulster at this time.[327]

One of the most ambitious of the early users of hydro-electricity was the Giant's Causeway tramway, built in 1883, the first electric railway in the United Kingdom. A watermill, most recently used for flax and before that as a paper mill, was equipped by the railway company with two turbines made by the New York firm of Allcott, each producing 45hp from a fall of 26 feet. They were replaced with more powerful turbines in 1900 and 1903. The example of the Giant's Causeway tramway was followed by the Bessbrook & Newry tramway, established in 1884; for this, a flour mill was converted to power generation with a Vortex turbine of 62hp, which lasted until 1921, when it was replaced. This tramway was an offshoot of the Bessbrook Spinning Company, built mainly to carry the firm's supplies and finished goods to and from the docks at Newry, while also offering a public passenger and freight service. It closed in 1948, when the spinning company suffered a downturn in trade, and had been served throughout by the hydro-electric power. The Giant's Causeway tramway lasted another two years. It, too, continued to find hydro-electric power very cost-effective, although an auxiliary gas-fired generator was used for the peaks of demand in the summer, when the river was at its lowest. Insufficient volume of passenger traffic forced the line's closure in 1950.[328]

The generation of hydro-electric power on a larger scale was also tied directly to industrial requirements. Aluminium-smelting works consumed large quantities of power, and cheap electricity became a pre-requisite. Hydro-electricity drew the smelters to the remoter parts of Britain. The first works was built at Foyers in the Scottish Highlands in 1896. It was followed by a much larger works at Kinlochleven in 1909, and a similar development in the Conway valley in north Wales.

The mines and quarries of north Wales were particularly keen to use electricity, for the cost of transporting coal made it an expensive fuel in these remote areas. Water power, by contrast, was abundant. The Llechwedd slate quarry was the first to install its own water turbine in 1890, a Vortex of 28hp. This was replaced in 1905 by the company's larger Pant yr Afon generating station, which had two Gilkes Pelton wheels running at 385rpm to drive two 250KW generators. This station was still in operation at the end of the century. The Croesor slate mine installed a impulse-type water turbine of 375hp in 1901–2. The Frongoch lead mine likewise went for hydro-electricity, installing a Pelton wheel in 1900.[329]

Demand from the mines attracted investment from public companies established to supply those that did not have their own generating capacity. The first was the Yale Electric Power Company, founded in 1899. This firm built a generating station at the Falls of Dolwen, near Blaenau Ffestiniog, which opened in 1900. It did not fully close until 1964, although it had been absorbed and altered by the surrounding pumped storage generating station built after the Second World War. The main customers for the company were the mines and quarries, but surplus power was also sold for the public supply at Blaenau.[330]

The North Wales Power and Traction Company was a much more ambitious undertaking and, perhaps as a consequence, took some time to get off the ground. One of its plans was to use some of the electricity it would generate to provide power for a light railway between Portmadoc and Caernarvon (the 'traction' part of the company's title). Initial proposals at the turn of the twentieth century failed to secure backing. The company that eventually emerged in 1903 secured statutory authority through the North Wales Power Act 1904, which gave it greater security in negotiating for wayleaves. A generating station was built at Cwm Dyli and opened in 1906. It had a generating capacity of 4,000KW produced by Ganz twin Pelton wheels each rated at 1,500hp. Its first major customer was to be the Aluminium Company, which opened a smelting plant at Dolgarrog in 1907. As a result of falling prices for aluminium, the firm was bankrupted in the next year.

It was resuscitated in 1909, under the direction of a local entrepreneur, H. J. Jack, and production restarted in 1910.[331]

After that hesitant and loss-making start, the North Wales Power & Traction Company became established as the major supplier of power in north Wales, serving by the 1920s an area of 4,000 square miles, extending into west Cheshire. In 1922 it formally abandoned its ideas of an electrified light railway, and sold its interests to a new Welsh Highland Railway company, which was controlled by H. J. Jack, who also had a leading interest in the neighbouring Ffestiniog and Snowdon Mountain Railways. The generating company now simplified its name to North Wales Power. It did not lose interest in transport entirely, proposing in the 1920s that its power should enable the railway line along the north Wales coast to be electrified. After the building of its new power station at Maentwrog, it did supply power to Crewe railway works.

North Wales Power developed three large hydro-electric power stations: the original one at Cwm Dyli, Dolgarrog by the aluminium smelter, and Maentwrog. These were the principal generating stations in north Wales and there were only three steam generators in operation in the district when the electricity-supply industry was nationalized in 1948. Hydro-electric power did not account for the whole of the local consumption, however, for additional supplies were drawn from Merseyside.

Dolgarrog power station underwent extensive expansion during the 1920s, with new Pelton wheel turbines and additional water supply being drawn from Llyn Eigiau and Llyn Coedty. The failure of a dam in 1925 after a summer of drought was a major setback, but the generating capacity was restored, and increased by 6.5MW in 1936. A further expansion in 1957 brought its total power capacity to 26.5MW. Maentwrog power station was opened in 1928, the first to draw its water supply from an artificial reservoir. It was built with three Boving horizontal impulse turbines and a fourth was added in 1934. When it opened, it was the second largest hydro-electricity scheme in Britain after Kinlochleven.[332]

A number of other hydro-electric power stations were built in the north Wales area in the early twentieth century, including a low-fall generating station built in 1913 on the River Dee for the city of Chester. It was designed by S. E. Britton, the city corporation's electrical engineer from 1904 to 1946, and at the time produced 40 per cent of the city's consumption. The old Dee Mills flour mill was converted to a generating station in this scheme. After the Chester scheme Britton went on to work on a power station for the North Wales County Asylum, in which a vertical-shaft turbine of 120hp was incorporated into a dam on the River Ystrad. Between 1917 and 1920 Britton worked on a greater scheme for sixteen low-fall power stations to be built along the Dee, with associated dams and weirs to control the flow of water. It was a part of his vision for the cheap supply of electricity to enable agriculture and rural industry to flourish in this remote part of the country, but it was not accepted by the Electricity Commissioners, who preferred the North Wales Power Company's plans.[333]

Developments in Scotland followed a similar pattern. The first large hydro-electric power stations were built to supply the aluminium plants at Foyers and Kinlochleven built by the British Aluminium Company. The smelter at Kinlochleven was a major development, and the power station of 24MW was the biggest in Britain for its time, although both were small by Canadian standards, where aluminium production had already developed. Another smelter was built at Fort William in 1929 and, following its expansion, a much larger power station was also built, eventually reaching a capacity of 85.7MW.[334]

It was not until the establishment of the national grid that Scottish hydro-electric power could expand, for local demand was limited, while aluminium remained almost the only industry to be attracted into the highland areas where the power was produced. The grid enabled power to be exported to areas of higher demand, and could exploit one of the main advantages of hydro-electricity: the ability to start generation quickly to meet peaks in demand. The Electricity Supply Act 1926 established the grid, and in the succeeding few years the network of transmission lines and substations was constructed, the final links being

Pitlochry dam and hydro-electric power station on the River Tummel, seen in 1954. A salmon ladder is to the left of the power station. (Museum of English Rural Life)

completed in 1933. Moves to create the grid prompted the construction of new hydro-electric power stations for the public supply. Falls of Clyde power station was built in 1926, followed by some large stations at Rannoch in 1930 and Tummel in 1933. The Galloway Water Power Company was established in 1931 with a scheme to build five power stations, completed in 1935–6. This scheme was built specifically to provide power at peaks of demand. By 1938 the generating capacity of British hydro-power stations for public supply had reached 986GWh, most of it in Scotland and amounting to about 4.3 per cent of the national total, compared with a negligible amount in 1918.[335]

There was considerable opposition to the expansion of hydro-electric power in Scotland from landed and sporting interests, who believed power generation would destroy highland wilderness, and from coal owners concerned about potential competition. Between them, these interests were strong enough to defeat a number of planned developments in the country, with six Bills rejected in parliament from 1929 to the Second World War. The wartime secretary of state for Scotland, Tom Johnston, appointed an investigatory committee under Lord Cooper to break the deadlock. Inspired by the Tennessee Valley Authority in America, its report recommended the transfer of responsibility for power generation in northern Scotland to a public body, the North of Scotland Hydro-Electric Board. It was a solution that suited the predispositions of the minister, as a Labour politician. The board was set up in 1943, retaining a strong measure of independence after the rest of the British power-generating industry was nationalized in 1948.[336]

Expansion schemes followed the creation of the new board, systematically exploiting the area's potential, and they were seen through despite some continued opposition. The Loch Lomond scheme required the construction of a buttress dam 168 feet high, the tallest then in Scotland, to raise the waters of Loch Sloy. The Tummel-Garry power stations, already developed by the Grampian Company, were extended, and developments were undertaken in Glen Affric, Conon Valley, Garry Moriston, and Shin. By 1960 total power generation by Scottish hydro-electricity had reached 875MW, ten times what the board had started with. There were thirty-one power stations in the Grampian region, while expansion of the Galloway scheme had resulted in six power

Hydro-Electricity

Fasnakyle dam, part of the Glen Affric power development. (Museum of English Rural Life)

stations, with maximum output capacity of 109MW, along the River Dee, fed by Loch Doon, Loch Ken, and the artificial reservoir of Clatteringshaws Loch.[337]

Hydro-electricity was looked upon as a force for economic growth in the Highlands. Direct employment in the industry has, however, been small – apart from the temporary influx of many thousands involved in the construction of the power stations – for the stations can almost run themselves. With little by way of new industry following the aluminium smelters into the region to use the power, most of what has been generated has been used, via the national grid, outside of the Highland zone. Although the rate of expansion slowed as the availability of sites diminished, the number of Scottish hydro-electric power stations had reached more than sixty at the beginning of the twenty-first century. Almost all are in the western half of the country, where the lakes and higher rainfall offer the greatest potential for development. The rainfall, highest in the winter, coincides with the season of greatest demand for power.

The development of public hydro-electric power schemes in Northern Ireland followed the pattern of growth from small and private beginnings. An old bleach green at Largy on the River Roe was converted in 1894 to provide power for the house at nearby Roe Park. After a year or so an American 'Little Giant' turbine replaced the waterwheel, the success of which led to the formation of the Limavady Electricity Supply Company to generate power for public supply. In 1918 hydro-electricity replaced the town's gaslights. By that time the company had upgraded its capacity with three new turbines. Further expansion followed in 1926, when another mill was taken on and converted to power generation. This company continued in business, to be taken over by the Electricity Board for Northern Ireland, unlike most of the other small-scale power schemes for townships, which had closed down during the 1930s.[338]

The potential of the Lower Bann to provide power on a large scale was recognized by the mid-nineteenth century, a time when only mills of modest scale were able to draw upon it. Greater interest arose as hydro-electricity became a reality. The River Erne, on the north-west coast of what was later the Irish Republic, was also tapped for electrical power. The owner of Ballyshannon sawmills converted a grain mill in 1908 into a generating station, at first for his own needs; later, the power station was expanded to supply the town of Ballyshannon and Bundoran a little further down the coast, a role it performed until 1950. Proposals were made for electricity generation on a larger scale by the Irish Hydro-Electric Syndicate Ltd in 1915, but the war put paid to them, and it was not until 1946 that the Electricity Supply Board developed a scheme for power stations on the river.[339]

Hydro-electric power brought with it a new demand for water storage. Developments in the technology of dam-building, especially the 'arched' dam with its curved profile, came to the aid of the constructors of large new reservoirs. The Blackwater Reservoir was the first major dammed reservoir of this era, built between 1905 and 1909

Tummel power station.

for the Kinlochleven power scheme in Scotland. It was the largest in Europe when it was built: 8 miles long, retained by a dam 90 feet high and 3,000 feet wide. At about the same time, dams were being built for power schemes in Wales, the largest being at Maentwrog. In undertaking its large-scale expansion of power generation after the Second World War, the North of Scotland Hydro-Electric Board built fifty dams and reservoirs, most of them of the concrete buttress type. In Wales four dams were built for hydro-electricity schemes in 1961–2, two of them for the Ffestiniog pumped storage scheme.[340]

TIDAL POWER

Capturing the power of the tides to generate electricity has been a long-standing dream. Tidal power generation works on the same principle as the tide mill – the capture of water on the incoming tide to use as it flows back out. Electricity generation using the tides has been on a larger scale than the tide mills, and the costs and complexities involved have meant that throughout the world very few projects have been built. One of the biggest has been at Rance in France, often held up as an exemplar of the success of this technology. A more typical experience, perhaps, is that of the Bay of Fundy in Canada, which has a large tidal range, and has been recognized as having great potential. Although a pilot project costing $57 million was built in the early 1980s, no further development was made.

There is great potential for tidal power in the United Kingdom. One recent estimate puts it at as much as 47.7 per cent of the total for Europe. There have been proposals for the construction of barrages since the mid-nineteenth century, affecting the Mersey, Humber and Severn estuaries, the Wash and Morecambe Bay, but none, so far, has been built. Interest has grown more recently, encouraged by government pronouncements about developing renewable energy in the early 1990s, and some proposals were worked into quite an advanced state. The Mersey estuary is reckoned to have potential power of 700MW; a Mersey Barrage Company was set up, and detailed design work undertaken, but so far nothing has come of it. Proposals for barrages across Strangford and Carlingford Loughs in Northern Ireland have not come to fruition either, although co-operation between the two governments in Ireland did result in the Lough Erne catchment being tapped between Belleek in County Fermanagh and Ballyshannon in County Donegal. Two large power stations were constructed, at Cliff and Catthaleen's Fall.[341]

One of the most ambitious proposals for generating electricity using tidal power has been the Severn Barrage. The idea of putting a barrage across the river has a long history. The first proposal was put forward back in 1849, although there was no thought then of using the high tidal range of 45 feet to produce electricity. That was first introduced with a report from the Ministry of Transport in 1920. Committees, investigations and reports have followed with almost monotonous regularity since then, with several sites being proposed. The Severn Barrage Committee, set up in 1978 under the chairmanship of Sir Hermann Bondi, recommended in 1981 that the barrage should stretch from near Cardiff to near Weston-super-Mare, and this has been the preferred option ever since. Another report, from the Severn Tidal Power Group, argued that 'In view of the economic and other benefits potentially available ... the building of an energy-generating barrage ... would be a great value and a permanent asset to the country.'[342]

Despite such strong endorsement, the schemes have so far always fallen by the wayside because of the high capital cost, uncertainties over the technology that expenditure will buy, and the politics involved. Many other interests would be affected by any hydro-electricity scheme: industry and commerce along the river, leisure and sporting interests, and environmental concerns. Environmental groups especially have opposed the construction of a barrage across the Severn, for it would destroy extensive tidal mudflats, home to many thousands of coastal birds, as well as fish and other wildlife. Towards the end of 2008 the Severn Barrage project resurfaced, with the government publishing a report on proposals in January 2009. A number of schemes were included in the proposals: the largest was the 10-mile barrage from Cardiff to Weston-super-Mare, with two smaller barrages and coastal lagoons at Fleming and Bridgwater Bay making up the shortlist of proposals. The Cardiff barrage power station would generate an estimated 8GW; the tidal lagoons could add another 1.6GW. The cost remains high – estimated at £14–15 billions for the Cardiff barrage, and the budget cuts of 2010 deferred the project once again.[343]

14 A Green Future for Water Power?

So rapid was the decline in the numbers of rural watermills, especially of corn mills after the 1880s, that by the 1890s writers were beginning to mourn their passing. Bennett and Elton, in their volume published in 1899, wrote of 'the once "busy mill" edged out of existence by the giant stride and power of the modern roller mill. Vast numbers of the venerable rustic structures have succumbed to ruin; many have fallen from their high estate as grinders of flour to that of makers of cattle food ... not a few, constantly appearing on the market, change hands at prices testifying but too palpably to their generally diminished value'. In the early twentieth century it became common for writers on country matters to lament 'watermills in decay', and describe the 'delightful hobby' of tracing the course of streams, 'camera in hand', to search out the old mills hidden away in odd corners.[344]

Moves to record and preserve mills began in the early years of the twentieth century. The first to be preserved – as early as 1904 – was the mill at Troswick in the Shetlands, where the horizontal wheel was restored by Gilbert Goudie.[345] However, the watermill never excited quite as much interest as the windmill. Perhaps this was because William Coles Finch was not the only one who found 'windmills to be more elusive than one would imagine, but the watermill is more so'.[346] While the Society for the Protection of Ancient Buildings, established in 1877, had set up a windmill section early on, it was not until

A mill in decay at Sawbridgeworth, Hertfordshire, in 1980.

A Green Future for Water Power?

Newnham Mill, Northamptonshire, being demolished in August 1954. (Museum of English Rural Life)

1946 that this was expanded to include watermills. Once involved, the society became very active and carried out a number of important surveys in the 1950s of working mills and disused mills that survived in one form or another in various states of repair.

Interest in the mills gathered pace after 1945. By the end of the century more than a hundred mills had been preserved in one way or another. Most of them are corn mills, but many others are from industry and mining. Several bodies have been involved: the Ministry of Public Building and Works (later English Heritage), local authorities, the National Trust, preservation societies and trusts, and some individual owners. Some preserved mills have been physically removed from their site to open-air museums. The Esgair Moel woollen mill from Breconshire, built in 1760 and worked until 1947, was given to the Welsh Folk Museum in 1950. It was moved to St Fagan's and restored to working order. The Castle Museum at York has its working corn mill. Not all of these preserved mills are able to use water power alone. Damage to weirs or loss of leats means that some have to use electricity to pump water into a short feeder channel to demonstrate their operation.

Cheddleton flint mill was one of the first industrial mills to be preserved, by a trust formed in 1967. It was followed by the industrial site at Abbeydale, near Sheffield, where the water-powered iron forges that had ceased working in 1933 were preserved in 1970. The Finch foundry at Sticklepath, Devon, has also been restored as a working museum. The National Trust has become the custodian of Quarry

Ixworth Mill, Suffolk, in a sorry-looking state in the mid-twentieth century. (Museum of English Rural Life)

A Green Future for Water Power?

Woodbridge tide mill was nearing the end of its working life when this photograph was taken. It closed in 1957 and has since been preserved in working order. (Museum of English Rural Life)

Bank Mill, Styal, and has recently restored its waterwheel, a rare survivor among the large industrial wheels; those at Catrine, Deanston and Belper have not survived.

Although restored to working order, many, perhaps most, of the preserved watermills are not able to use water as their main source of power. The wheels are turned occasionally, on special open days or to replenish supplies of home-milled flour, perhaps, while for most purposes mains electricity provides the power. Doubts about the reliability of the flow of water, restrictions on the amount of water that can be extracted or the need to conserve the wheels and gearing are the reasons, but it all means that the preservation of the mills hardly constitutes a revival of the use of water power.[347]

The rate at which mills were closing, to be turned into flats and houses, restaurants and shops, or demolished, was the driving force behind the preservation movement. Those buildings that were converted to other uses received treatment of varying degrees of

Water pours into the relatively new iron overshot wheel at the preserved Finch foundry.

There is little now to indicate that this house at Grindleford, Derbyshire, was once a watermill, although the millstream still flows under the bridge.

Mapledurham Mill on the River Thames occupies a site recorded in Domesday Book. It is now working again producing flour regularly and open to the public.

sympathy. Arkwright's Masson Mills at Cromford have become a shopping centre. The Dunkirk Mills at Nailsworth, Gloucestershire, were converted to flats in 1999, with the wheels being restored and a little museum included in the development.[348]

At the same time there was a remarkable number of instances of waterwheels and turbines continuing to provide power into the 1960s, grinding corn, driving textile mills, generating electricity. There was hardly a break between the long-time use of water power and the renewal of interest in 'alternative' fuel in the late twentieth century. This revival of working water power has involved mainly the two areas of food and electricity generation.

Since the late 1970s the growth of the whole-food movement has led to increased demand for stone-ground flour, and to a minor revival in the fortunes of windmills and watermills. Some preserved mills had already been restored to working order, grinding local grain for sale to visitors, and their numbers have increased with the growing interest in whole foods. Meanwhile a few private owners have restored mills as fully or partly commercial operations. Bacheldre Mill near Montgomery in Wales has been restored as a commercial mill, specializing in the production of spelt flour.

Mapledurham Mill, on the River Thames not far from Reading, is owned by the neighbouring agricultural estate. There had been a mill on the site since at least the record in Domesday Book, but it finally ceased grinding in the 1940s and fell into disrepair. In the 1970s it was restored to working order – the funds deriving from income earned when it was used as a film location – and since 1977 it has built up a good local trade for the 5 tons of flour it can produce each week. By the end of the century the Traditional Corn Millers Guild, founded as recently as 1987, had thirty members who were operators of watermills and windmills. It is a far cry from the thousands of water corn mills that used to adorn the landscape, but it is a small indication that water power is not finished yet.

Water power can also generate 'green' electricity, as it has done since the late nineteenth century. Although large-scale hydro-electricity generation has grown immensely since 1945, it accounts for a very small part of the UK's total: about 2 per cent of generating capacity and 1 per cent of output. By the 1970s the opportunities for building further power stations of this type had become limited, with few sites available in the highland regions where they have been concentrated. With growing interest in renewable power, investment in some of the remaining sites has begun. Glendoe, near Fort Augustus, was opened in 2009, the first large-scale hydro-electric power station built in Scotland for fifty years.[349]

Power can also be generated on a smaller scale. Power generation of less than 5MW is now officially counted as 'small hydro', while less than 100KW is 'micro-hydro'. Generation on this scale has always been a possibility for domestic use on farms, by small businesses and sometimes for public supply. During the course of the twentieth century a number of water-power sites were converted to electricity generation. At Belper, the Strutts' South Mill was demolished and replaced by a generating station with water turbines. In 1953 the owner of the site of Lowwood gunpowder mills, Furness, reconstructed the weir and leat to provide for two 150KW hydro-electric generators that supplied power to the national grid.[350]

While many owners were turning their mills over to electricity generation, general interest was limited during the first three-quarters of the twentieth century. An article published in 1946, 'Generating electricity on the farm', reflected the common view; while it mentioned using the almost free energy from streams, it also assumed that farmers would be buying oil-powered generators.[351] Officialdom thought only in terms of the large-scale public hydro-electricity generators: there was no concept of smaller producers supplying the national grid. The licensing regime of the Water Resources Act 1963 had the effect of restricting small-scale use of water power and stifled development of small-scale hydro-electricity generation; it also led to the disappearance of the wind pumps that had once been a common sight on the farm. Some relaxation of the Act's conditions was only achieved as a result of a campaign by the National Association of Water Power Users.[352]

More recently, the use of water power for electricity generation has advanced up the political agenda, and interest in smaller-scale applications has grown, especially since the 1980s. The Non-Fossil Fuel Obligations imposed by government encouraged this movement. With ever more ambitious

The reservoir at Glendoe in the midst of a Highland winter. (Scottish and Southern Electricity plc)

targets for the proportion of power to be generated from renewable sources being declared by the European Union, and, at least as important, prices for oil and gas rising, the search has intensified for alternative, green means of generating power. The introduction of the Renewables Obligation in 2002, a levy on electricity bills to subsidize renewable energy, was another indication of government support. By 2007 the total sums raised through the levy had reached more than £1 billion a year, of which about 12 per cent went to hydro-electricity generation, while wind power received 40 per cent. It might seem odd that so little prominence has been accorded to water power, while wind farms have attracted enormous attention compared with the amount of electricity they generate. In the background, however, the number of projects to harness water power has been growing. There has been sponsorship of research and pilot projects from both the British government and the European Union. Some of these have been test-beds for new technologies for micro-hydro-electric power, including low-cost inflatable rubber weirs and low-head generation. The results of these programmes have been variable. One of the European programmes, started in 1990, involved 154 projects across the European Union, fourteen of them from the United Kingdom. Only four of the British ones reached an advanced stage, however.[353]

By the 1990s small-scale hydro-electricity projects were being developed on a number of private estates, which are often well placed to undertake them. One of the first was built by Paul Bromley in 1991 on the 20-acre estate he owned in Cheshire. It was on the very small scale, with a 8-foot diameter by 3-foot wide vertical wheel generating 1½KW. The power helped with Bromley's domestic needs, but the installation was mainly used for educational purposes through his Pedley Wood Conservation Trust.[354] Buckfast Abbey, Devon, installed water-powered electricity generation in 1997.[355] The Royal estate at Balmoral has its own hydro-electric generator. It uses a mill in which a turbine was installed in the 1920s to produce electric light, but which was used in the 1950s to drive the estate sawmill. Since then it has lain out of use, and the turbine has recently been overhauled to generate about 1MW of electricity, enough to make the estate more than self-sufficient. Windsor Castle is planning to generate 30 per cent of its power consumption from a hydro-electric plant to be built on an island in the River Thames.[356]

A number of private owners of watermills that have been converted to dwellings have started bringing them back to life as power generators, and some new restorations of derelict mills have incorporated hydro-electricity generation. Dandridge's Mill at East Hanney, Oxfordshire, a mill that at different stages in its history produced silk and flour, has been converted to apartments incorporating power generated from its millstream.[357]

The Elan Valley electricity scheme in mid-Wales is one of the first of a generation of small-scale public hydro-electricity producers. Started in 1998, it is unusual in using existing infrastructure of dams and reservoirs built early in the twentieth century to feed Birmingham's water supply. The idea of using this to generate power was not new – there was already one small power station at Caban Coch. In the new scheme, its turbine was renewed and four new power stations added. The scale has remained small and the total capacity is only 4.2MW.[358]

Many recent developments of projects for public supply have been community-based social enterprises. Their objective has been to generate enough power for their own local needs and then to plough back into community projects any earnings from selling surpluses to the national grid. Torrs Hydro in Derbyshire was the first community-owned hydro-electricity generator to start production, in the autumn of 2008. At the weir that once supplied Torr Mill, a textile mill on the River Goyt that burned down in 1912, an Archimedean screw has been installed to generate electricity. The installation was undertaken by Water Power Enterprises, another social enterprise company based in Todmorden and set up to encourage local initiatives, with ambitions for about two dozen small-hydro power plants to be in operation by 2015. A number of other community schemes are planned, including one at Settle, a mill town in the Yorkshire Dales, Stockport Hydro Ltd, launched in March 2010 to develop more sites on the

River Goyt, and Esk Energy, which plans to generate power at the Ruswarp weir on the River Esk, North Yorkshire.[359] These projects nearly all use existing water-power sites. The weirs and other works are already there, requiring at most repair and modification; moreover, use of an existing site is likely to ease the owner's passage through the bureaucracy of water-extraction licences.

The water turbines that have been the main means of generating electricity are not always suited to small-scale, low-fall sites. The types of turbine that can work with a low head, the Kaplan and the cross-flow, tend to be expensive, needing complex engineering to install them, and an efficient filter to keep debris in the stream from fouling the wheel. The preferred technology for small-scale projects has become the 'reverse Archimedean screw', so called because the screw as originally devised about 2,000 years ago was a lifting mechanism using power from another source. Research in the 1990s showed that it could become an effective generator of power through the action of water descending on its vanes causing it to revolve in the opposite direction. The advantage of the Archimedean screw for small-scale power generation is that it can work with a modest fall of water and, sited alongside a weir, requires little attention. It is also claimed to have minimal effect on fish and fowl in the river.

There has also been revived interest in the vertical waterwheel as a means of generating power on the small scale. Overshot wheels, some new, some restored, have been used. Hydrowatt, a German company, has also promoted the qualities of the Zuppinger type of undershot wheel, one of which has been installed at the offices of Ramblers Holidays Ltd at Lemsford Mill in Hertfordshire. This is a type of wheel new to the British Isles, although it was developed by Walter Zuppinger as long ago as 1883. It was one of a number of developments made from the Poncelet wheel undertaken by European engineers. In 1858 Alphonse Sagebien modified the breast-shot wheel with inclined blades to reduce the impact of descending water. Sagebien wheels were very efficient, but none were built in Britain. Zuppinger modified the Sagebien wheel by giving the floats a gentle curvature.[360]

Despite the advantages of Archimedean screws or efficient wheels, the cost of installing small-scale generating plant is not insignificant. The Archimedean screw at Torrs Mill cost £300,000 and will produce 250,000KWh per year. It does compare favourably with wind power, however: twelve wind turbines 12 metres high would cost the same to construct and generate about half the power.

The optimists believe there is potential for up to 1,000MW of power to be generated by small plant

Torrs Hydro's micro-power generator at New Mills. The reverse Archimedean screw is enclosed in the green safety cage, receiving water diverted by the weir just as the former mill did.

at water-power sites. By 2002 the capacity for small-scale hydro-electric power generation in the United Kingdom had reached 70MW, with an annual output of about 200GWh. This is only 4 per cent of the total hydro-electricity production, and a tiny fraction of all the nation's electricity. Nevertheless, it has been growing at a healthy rate, about 6 per cent a year, since 1990.[361]

Some still continue to dream of tidal power. The main Severn Barrage would generate 5 per cent of the United Kingdom's power consumption, according to the government's advisors. Other schemes on the shortlist of proposals would produce less. The more ambitious hopes are that tidal energy could provide as much as 20 per cent of the nation's energy needs. The Carbon Trust, one of the government agencies behind such ambitions, was to invest £1m in research into reducing the cost of tidal power generation using 'giant turbines'. Research conducted at the University of Michigan seemed to be opening up the prospect of increasing the amount of power generated from tidal schemes, by enabling more estuaries to be used. Most water turbines need a current flowing at 5 or 6 knots. Only a small proportion of the world's estuaries reach that speed, but the Michigan turbine can work at less than 1 knot, offering the potential for putting more rivers to work.[362]

There is another form of power generation from the sea. The tidal current generator uses the horizontal flow of the current to turn the turbine, unlike the estuarine barrage, which relies on the rise and fall of the tide. A prototype was tested in 1994 near Fort William. Another tidal power generator was tested off the Devon coast near Lynmouth by a company called Marine Current Turbines, and in 2003 they installed a tidal turbine between Foreland Ledge and Foreland Point, a mile off the shore. The turbine is fitted to a steel pile driven into the sea bed, with the top of the pile protruding only a short way above the surface of the waves. The current acts upon what looks like a submerged wind turbine. On the 300KW prototype it was possible to make use of the tidal flow in one direction only. After working successfully for more than three years, the company has plans to develop the technology on a larger scale and expand the site at Lynmouth. Another type of tidal generator, called the Stingray, was tested in Yell Sound, Isles of Shetland, in 2002–3.[363]

Aquamarine Power's Oyster hydro-electric wave energy converter in operation. (aquamarinepower.com)

A Green Future for Water Power?

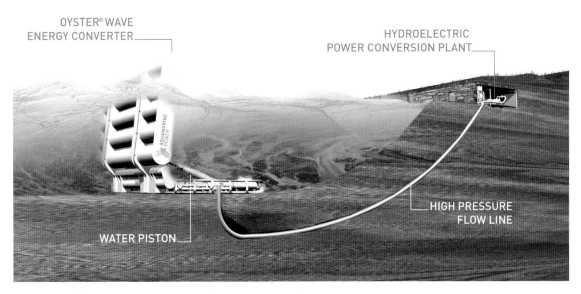

The workings of the Oyster wave energy converter. (aquamarinepower.com)

Wave energy is another source of power that has recently come to the fore. It attempts to tap the power in sea waves either offshore or as they hit the land. Some estimates put the potential energy from wave power around the United Kingdom at 7GW. The concept of capturing this power is an old one, with ideas being put forward at least 200 years ago, but workable technology only began to be developed in the 1970s, spurred on by the energy crisis of 1973. A number of governments, the British included, had taken an interest, but, with the relieving of that crisis, research into wave technology diminished. The next round of interest in renewable energy came in the 1990s and research has continued, mainly on a modest scale but making slow and steady progress. Among these development projects is one on the Isle of Islay, which enjoys the low tidal range preferred for this type of power. An oscillating water-column wave energy converter has been working here for some years.[364]

At the beginning of 2010 a major boost to the prospects for the generation of both wave power and tidal power came with the award by the Crown Estate of development licences for sites off the Orkney Isles and the coast of Sutherland and Caithness. The intention is that these sites should yield as much as 1.2GW of power.[365]

15 Conclusion

The waterwheel at Coley Farm, Staffordshire.

Water power has had a remarkably complex history, interwoven within many aspects of economy and society, from the medieval estate to the Industrial Revolution and on to hydro-electricity. Far from being a technology of an earlier age inevitably superseded by steam, diesel and electricity, water power has proved itself to be immensely adaptable, co-existing with these other sources of power. The complexities of the story of water power are to be seen in its impact on the landscape, in the uniqueness of almost all sites exploited for it, and in the way they have been developed and re-used for different purposes.

The problem with water power in the British Isles has always been that it is not universally available. As demand for power has grown, so it has been left behind as a major provider since the mid-nineteenth century. However, the continual search for new sources of energy means that, after 2,000 years of development, water power remains an active force. Its story is far from over.

Notes

1. Lewis Mumford, *Technics and Civilization* (1934). R. J. Forbes, *Studies in Ancient Technology* (1955).
2. G. N. von Tunzelmann, *Steam Power and British Industrialization to 1860* (1978) pp. 1–7 discusses this point.
3. Peter Lord, *Portrait of the River Trent* (1968) p. 89.
4. Quotation from George Eliot, *The Mill on the Floss* (1860) ch. 1.
5. Quoted in Kenneth Clark, *Civilisation* (1969) p. 281. P. A. Rahtz, 'Medieval Milling', in D. W. Crossley, ed., *Medieval Industry* (1981) p. 1.
6. Uvedale Price, *Essays in the Picturesque* (1810) v. 1, p. 55. Richard Hayman, 'Artists' Impressions of Aberdulais Mill', *Industrial Archaeology Review*, v. 9 (1986–7) pp. 155–66.
7. Edmund Blunden, 'Country Childhood', in Simon Nowell-Smith, ed., *Edwardian England* (1964) pp. 556–7.
8. M. J. T. Lewis, *Millstone and Hammer: The origins of water power* (1997) p. 18. J. P. Oleson, *Greek and Roman Mechanical Water-lifting Devices: The history of a technology* (1984) pp. 291–301, 325ff.
9. M. J. T. Lewis, *Millstone and Hammer* (1997) pp. 20ff, 49–62. N. A. F. Smith, 'The Origins of Water Power: A problem of evidence and expectations', *Transactions of the Newcomen Society*, v. 55 (1983–4) pp. 67–84. Orjan Wikander, 'Archaeological Evidence for Early Water Mills – an interim report', *History of Technology*, v. 10 (1985) pp. 151–80.
10. Richard Bennett and John Elton, *History of Corn Milling*, vol. 2 (1899) pp. 6–30. N. A. F. Smith, 'The Origins of Water Power: a problem of evidence and expectations', *Transactions of the Newcomen Society*, v. 55 (1983–4) pp. 67–84. Orjan Wikander, 'Archaeological Evidence for Early Water Mills – an interim report', *History of Technology*, v. 10 (1985) p. 164. Reynolds, *Stronger than a Hundred Men*, pp. 14–26.
11. J. G. Landels, *Engineering in the Ancient World* (1978) p. 17.
12. George Basalla, *The Evolution of Technology* (1988) pp. 144ff. Orjan Wikander, 'Archaeological Evidence for Early Water Mills – an interim report', *History of Technology*, v. 10 (1985) p. 155. Reynolds, *Stronger than a Hundred Men*, pp. 30–44.
13. R. J. Spain, 'The Second-century Romano-British Corn Mill at Ickham, Kent', *History of Technology*, v. 9 (1984) pp. 143–80. R. J. Spain, 'Romano-British Watermills', *Archaeologia Cantiana*, v. 100 (1984) pp. 101–28.
14. Reynolds, *Stronger than a Hundred Men*, pp. 48–51. Maurice Beresford and John Hurst, *Wharram Percy Deserted Medieval Village* (1990) p. 66.
15. Colin Rynne, 'The Introduction of the Vertical Watermill into Ireland: some recent archaeological evidence', *Medieval Archaeology*, v. 33 (1989) pp. 21–31. Thomas McErlean and Norman Crothers, *Harnessing the Tides: the early medieval tide mills at Nendrum Monastery, Strangford Lough* (2007) pp. 10–11.
16. C. J. Bond, 'The Reconstruction of the Medieval Landscape: the estates of Abingdon Abbey', *Landscape History*, v. 1 (1979) p. 69.
17. Margaret T. Hodgen, 'Domesday Water Mills', *Antiquity*, v. 13 (1939) p. 262.
18. H. C. Darby, *Domesday England* (1986 edn) pp. 271–2, 361.
19. John K. Harrison, *Eight Centuries of Milling in North East Yorkshire* (2nd edn 2008) pp. 2–4.
20. H. C. Darby, *Domesday England* (1986 edn) p. 361.
21. Richard Holt, 'Medieval England's water-related technologies', *Working with Water in Medieval Europe: Technology and resource use*, ed. Paolo Squatriti (2000), p. 60–1. Margaret T. Hodgen, 'Domesday Water Mills', *Antiquity*, v. 13 (1939) pp. 261–79. D. Dilworth, *The Tame Mills of Staffordshire* (1976) p. 11.
22. Richard Holt, *The Mills of Medieval England* (1988) pp. 159ff, 178–80. Margaret Yates, *Town and Country in Western Berkshire c1327–c1600: Social and economic change* (2007) p. 35.
23. Enid Gauldie, *The Scottish Country Miller 1700–1900: A history of water-powered meal milling in Scotland* (1981) p. 44.
24. Richard Holt, *The Mills of Medieval England* (1988) pp. 49–50. Dave Gregory and Michael Jones, *Norwell Mills* (Norwell Heritage Booklet 3, 2009) p. 9.

25 Maurice Beresford and John Hurst, *Wharram Percy Deserted Medieval Village* (1990) p. 67. Richard Holt, *The Mills of Medieval England* (1988) pp. 40–5. Rosamond Faith, 'The "Great Rumour" of 1377 and Peasant Ideology', in *The English Rising of 1381*, ed., R. H. Hilton and T. H. Aston (1984) pp. 65–6.

26 Richard Bennett and John Elton, *History of Corn Milling*, v. 2 (1899) pp. 108–13, 117. Richard Holt, *The Mills of Medieval England* (1988) pp. 92–4, 96–7.

27 Richard Holt, *The Mills of Medieval England* (1988) pp. 97–9. Gillian Clark, *Down by the River: The Thames and Kennet in Reading* (2009) pp. 57–8.

28 H. D. Gribbon, *The History of Water Power in Ulster* (1969) p. 43. Enid Gauldie, *The Scottish Country Miller*, p. 45. John Farey, *General View of the Agriculture of Derbyshire* (1806) v. 2, p. 492. W. L. Norman, 'The Wakefield Soke Mills to 1853', *Industrial Archaeology*, v. 7 (1970) pp. 176–83.

29 Richard Holt, 'Medieval England's water-related technologies', *Working with Water in Medieval Europe: Technology and resource use*, ed. Paolo Squatriti (2000), p. 66.

30 M. J. T. Lewis, *Millstone and Hammer* (1997).

31 Timothy E. Powell, 'The Disappearance of Horizontal Watermills from Medieval Ireland', *Transactions of the Newcomen Society*, v. 66 (1994–5) p. 220.

32 John Shaw, *Water Power in Scotland 1550–1870* (1984). Richard Holt, *The Mills of Medieval England* (1988) pp. 120–2. Enid Gauldie, *The Scottish Country Miller* (1981) pp. 115–16.

33 Alistair G. Thomson, *The Paper Industry in Scotland 1590–1861* (1974) p. 20. Enid Gauldie, *The Scottish Country Miller* (1981) pp. 116–18.

34 John Langdon, *Mills in the Medieval Economy* (2004) pp. 84–8. Reynolds, *Stronger than a Hundred Men*, pp. 97–103, 165. John Fitzherbert, *Surveyinge* (1539) p. 92.

35 John K. Harrison, *Eight Centuries of Milling in North East Yorkshire* (2nd edn 2008) pp. 67–76.

36 D. Dilworth, *The Tame Mills of Staffordshire* (1976) pp. 8–9. Rex Wailes, 'Suffolk Watermills', *Transactions of the Newcomen Society*, v. 37 (1964–5) p. 102. H. D. Gribbon, *The History of Water Power in Ulster* (1969) p. 20.

37 N. A. F. Smith, 'The Origins of Water Power: A problem of evidence and expectations', *Transactions of the Newcomen Society*, v. 55 (1983–4) p. 70.

38 Reynolds, *Stronger than a Hundred Men*, pp. 172–81.

39 Reynolds, *Stronger than a Hundred Men*, pp. 172–8.

40 M. J. Orbell, 'The Corn Milling Industry in the Industrial Revolution', PhD thesis, University of Nottingham (1977) pp. 291–2.

41 John M. Steane, *The Archaeology of Medieval England and Wales* (1984) p. 169. Jennifer Tann, 'Some Problems of Water Power: A study of mill siting', *Transactions of the Bristol and Gloucestershire Archaeological Society*, v. 84 (1965) p. 53.

42 Josef Sisitka, 'Floating mills in London: An historical survey', unpublished BA dissertation, University of Reading (1992).

43 Reynolds, *Stronger than a Hundred Men*, p. 62. Norman Smith, *A History of Dams* (1971) pp. 164–5.

44 Donald Hill, *A History of Engineering in Classical and Medieval Times* (1984) pp. 47–9.

45 Martin Watts, *Water and Wind Power*, p. 63.

46 Harrison, *Eight Centuries of Milling*, pp. 18–21.

47 Henry Cleere and David Crossley, *The Iron Industry of the Weald* (1985) pp. 235–6.

48 John Langdon, *Mills in the Medieval Economy* (2004) pp. 88–9.

49 Margaret T. Hodgen, 'Domesday Water Mills', *Antiquity*, v. 13 (1939) p. 266. John Langdon, *Mills in the Medieval Economy* (2004) pp. 80–1.

50 Frank Nixon, *The Industrial Archaeology of Derbyshire* (1969) pp. 99–100. Reynolds, *Stronger than a Hundred Men*, pp. 130–2.

51 Richard Holt, 'Medieval England's water-related technologies', *Working with Water in Medieval Europe: Technology and resource use*, ed. Paolo Squatriti (2000), pp. 65–6, 95–7. Geoffrey Binnie, 'The Evolution of British Dams', *Transactions of the Newcomen Society*, v. 47 (1974–6) p. 207. David Luckhurst, *Monastic Watermills* (nd) p. 12.

52 Richard Bennett and John Elton, *History of Corn Milling*, v. 2 (1899) pp. 181–6.

53 Peter Lord, *Portrait of the River Trent* (1968) pp. 160–1.

54 Charles Hadfield, *The Canal Age* (1971 edn) pp. 2–4. Gillian Clark, *Down by the River: The Thames and Kennet in Reading* (2009) p. 29.

55 Harrison, *Eight Centuries of Milling*, p. 21. John Farey, *General View of the Agriculture of Derbyshire* (1806) v. 2, p. 489.

56 Enid Gauldie, *The Scottish Country Miller*, pp. 104–5.

57 C. J. Bond, 'The Reconstruction of the Medieval Landscape: the estates of Abingdon Abbey', *Landscape History*, v. 1 (1979) p. 69.

58 Giles and Goodall, *Yorkshire Textile Mills*, p. 130.

59 Hadrian Cook, 'A Tale of Two Catchments: water management and quality in the Wandle and Tillingbourne 1600–1990', *Southern History*, v. 30 (2008) p. 89.

60 R. S. Fitton, *The Arkwrights, Spinners of Fortune* (1989) pp. 55–6.

61 R. L. Hills, *Power in the Industrial Revolution* (1970) pp. 103–4, 108–9. Jennifer Tann, 'Some Problems of Water Power: A study of mill siting', *Transactions of the Bristol and Gloucestershire Archaeological Society*, v. 84 (1965) p. 75.

62 Reynolds, *Stronger than a Hundred Men*, pp. 18, 43–4.
63 See H. D. Gribbon, *The History of Water Power in Ulster* (1969) p. 25.
64 Vitruvius, *On Architecture*, v. 2, 304–7, quoted in Langdon, *Mills in the Medieval Economy*, p. 94.
65 Langdon, *Mills in the Medieval Economy*, p. 95.
66 Richard Holt, 'Medieval England's water-related technologies', *Working with Water in Medieval Europe: Technology and resource use*, ed. Paolo Squatriti (2000), p. 67.
67 Watts, *Water and Wind Power*, p. 40. Harrison, *Eight Centuries of Milling*, pp. 59–64.
68 Watts, pp. 11, 22.
69 Tann, 'The Employment of Power', p. 203.
70 Tann, 'The Employment of Power', p. 203.
71 W. Fairbairn, *Mills and Millwork*, v. 1.
72 Langdon, *Mills in the Medieval Economy*, pp. 103–7.
73 Watts pp. 30–1.
74 Alan H. Graham, 'The Old Malthouse, Abbotsbury, Dorset: the medieval watermill of the Benedictine Abbey', *Proceedings of the Dorset Natural History and Archaeological Society*, v. 108 (1986) pp. 103–25.
75 Thomas McErlean and Norman Crothers, *Harnessing the Tides: The early medieval tide mills at Nendrum Monastery, Strangford Lough* (2007). Leslie Syson, *British Water-mills* (1965) p. 92. John M. Steane, *The Archaeology of Medieval England and Wales* (1984) p. 170.
76 H. D. Gribbon, *The History of Water Power in Ulster* (1969) pp. 51–2.
77 Rex Wailes, *Tide Mills* (1938). Martin Watts p. 56.
78 Martin Watts, p. 107. *Milling* (March 1979). E. M. Gardner, *The Three Mills: tide mills part three* (1957).
79 Martin Watts, p.88. Leslie Syson, pp. 93–4.
80 R. Shorland-Ball, 'Worsborough Corn Mill, South Yorkshire', *Industrial Archaeology Review*, v. 2 (1978) pp. 248–51.
81 Josef Sisitka, 'Floating mills in London: An historical survey', *Industrial Archaeology Review*, v. 19 (1997) pp. 21–30.
82 Daniel Defoe, *The Complete English Tradesman* (1745) pp. 144–5, 291, 298–9. M. J. Orbell, 'The Corn Milling Industry in the Industrial Revolution', PhD thesis, University of Nottingham (1977) pp. 58–9, 63. Gillian Clark, *Down by the River: the Thames and Kennet in Reading* (2009) p. 20.
83 Orbell, *The Corn Milling Industry*, pp. 60–1.
84 Orbell, *The Corn Milling Industry*
85 For example, Dave Gregory and Michael Jones, *Norwell Mills* (Norwell Heritage Booklet 3, 2009) pp. 10–11.
86 Richard Holt, 'Medieval England's water-related technologies', *Working with Water in Medieval Europe: Technology and resource use*, ed. Paolo Squatriti (2000), p. 76–9. Lynn White Jr, *Medieval Technology and Social Change* (1962) pp. 79–134. E. M. Carus-Wilson, 'An Industrial Revolution of the thirteenth century', *Economic History Review*, v. 7 (1941) pp. 39–60.
87 R. J. Spain, 'The Second-century Romano-British Corn Mill at Ickham, Kent', *History of Technology*, v. 9 (1984) pp. 143–80. Margaret Yates, *Town and Country in Western Berkshire c1327–c1600: Social and economic change* (2007) pp. 105, 204–5.
88 Langdon, *Mills in the Medieval Economy*, pp. 40–7.
89 Hadrian Cook, 'A Tale of Two Catchments: water management and quality in the Wandle and Tillingbourne 1600–1990', *Southern History*, v. 30 (2008) pp. 89, 91–2. P. F. Brandon, 'Land, Technology and Water Management in the Tillingbourne Valley, Surrey 1560–1760', *Southern History*, v. 6 (1984) pp. 75–6.
90 J. L. Bolton, *The Medieval English Economy 1150–1500* (1980) pp. 157–8. A. Rupert Hall and N. C. Russell, 'What about the Fulling Mill?' *History of Technology*, v. 6 (1981) pp. 113–19. Margaret Yates, *Town and Country in Western Berkshire c1327–c1600: Social and economic change* (2007) pp. 87–90.
91 Marilyn Palmer and Peter Neaverson, *The Textile Industry of South-west England: A social archaeology* (2005) pp. 32–3.
92 E. M. Carus-Wilson, 'An Industrial Revolution of the thirteenth century', *Economic History Review*, v. 7 (1941) pp. 39–60. H. C. Darby, ed., *A New Historical Geography of England before 1600* (1976) pp. 277–9. J. L. Bolton, *The Medieval English Economy 1150–1500* (1980) pp. 157–8. J. G. Jenkins, ed., *The Wool Textile Industry in Great Britain* (1972) pp. 202–3, 282–4.
93 Peter Allen, 'How Paper Making Came to Colthrop', *Berkshire Old and New*, no. 4 (1987), pp. 12–13.
94 W. B. Crump and G. Ghorbal, *History of the Huddersfield Woollen Industry* (1935) pp. 43–50.
95 D. T. Jenkins, 'Early Factory Development in the West Riding of Yorkshire 1770–1800', *Textile History and Economic History: Essays in honour of Miss Julia de Lacy Mann*, ed. N. B. Harte and K. G. Ponting (1973) p. 255. J. G. Jenkins, ed., *The Wool Textile Industry in Great Britain* (1972) p. 27.
96 H. W. Strong, *Industries of North Devon* (1889) pp. 137–8.
97 Jennifer Tann, *Gloucestershire Woollen Mills* (1967) p. 22.
98 G. D. Ramsey, *The Wiltshire Woollen Industry during the Seventeenth Century* (1943) pp. 18–20.
99 Anthony Calladine, 'Lombe's Mill: An exercise in reconstruction', *Industrial Archaeology Review*, v. 16 (1993) pp. 82–99. Daniel Defoe, *A Tour through the Whole Island of Great Britain* (1724–6, 1974 edn) v. 2, pp. 156–7.
100 Anthony Calladine and Jean Fricker, *East Cheshire Textile Mills* (1993) pp. 9–10, 33–4, 57–8.

101 Museum of English Rural Life, Twyford Mill notes.
102 Martin Watts, p. 94.
103 H. R. Schubert, *History of the British Iron and Steel Industry from c450BC to AD1775* (1957) p. 141.
104 G. G. Astill, *A Medieval Industrial Complex and its Landscape: The metalworking watermills and workshops of Bordesley Abbey* (1993). H. R. Schubert, *History of the British Iron and Steel Industry from c450BC to AD1775* (1957) p. 134.
105 D. W. Crossley, 'Medieval Iron Smelting', in D. W. Crossley, ed., *Medieval Industry* (1981) pp. 35–7. Brian G. Awty, 'The Cumbrian Bloomery Forge in the Seventeenth Century and Forge Equipment in the Charcoal Iron Industry', *Transactions of the Newcomen Society*, v. 51 (1979–80) pp. 25–41.
106 H. C. Darby, ed., *A New Historical Geography of England before 1600* (1976) p. 281. H. D. Gribbon, *The History of Water Power in Ulster* (1969) pp. 70ff.
107 T. A. P. Greeves, 'The Archaeological Potential of the Devon Tin Industry', in D. W. Crossley, ed., *Medieval Industry* (1981) pp. 85–95. Arthur Raistrick, *Industrial Archaeology* (1973) p. 28.
108 Langdon, *Mills in the Medieval Economy*, pp. 40–7.
109 A. E. Musson, *The Growth of British Industry* (1978) pp. 37–8. Daniel Defoe, *The Complete English Tradesman* (1745) pp. 298–9.
110 Richard Holt, 'Medieval England's water-related technologies', *Working with Water in Medieval Europe: Technology and resource use*, ed. Paolo Squatriti (2000), pp. 72–5.
111 D. Dilworth, *The Tame Mills of Staffordshire* (1976) p. 12. R. A. Pelham, 'The Water Power Crisis in Birmingham in the Eighteenth Century', *University of Birmingham Historical Journal*, v. 9 (1961) pp. 67–9, 79.
112 John G. Rollins, *The Needle Mills* (1970).
113 Martin Watts, p. 64.
114 J. D. Marshall and John K. Walton, *The Lake Counties from 1830 to the mid-twentieth century* (1981) p. 12.
115 W. Branch Johnson, *Industrial Archaeology of Hertfordshire* (1970) p. 55.
116 R. L. Hills, *Papermaking in Britain 1488–1988* (1988) pp. 50, 53.
117 Brian G. Awty, 'The Cumbrian Bloomery Forge in the Seventeenth Century and Forge Equipment in the Charcoal Iron Industry', *Transactions of the Newcomen Society*, v. 51 (1979–80) pp. 25–6.
118 A. H. Shorter, *Paper-making in the British Isles: An historical and geographical study* (1971) p. 32. D. C. Coleman, *The British Paper Industry, 1495–1860* (1958) p. 163.
119 Victoria County History, Hampshire, v. 5 (1912), pp. 489–90. C. Vancouver, *General View of the Agriculture of Hampshire* (1813) pp. 402–4. Pigot, *Directory*, 1830. Kelly's Directory, Hampshire, 1890, 1923. A. H. Shorter, *Paper Making in the British Isles* (1971) pp. 104, 120, 124, 134.
120 R. L. Hills, *Papermaking in Britain 1488–1988* (1988) pp. 56–7. H. D. Gribbon, *The History of Water Power in Ulster* (1969) pp. 130–1.
121 D. Dilworth, *The Tame Mills of Staffordshire* (1976) p. 11. Neil Wright, *Lincolnshire Towns and Industry 1700–1914* (1982) pp. 27, 78. Joan Thirsk, *Alternative Agriculture* (1997) pp. 74–6. Martin Watts, pp. 56–7.
122 P. F. Brandon, 'Land, Technology and Water Management in the Tillingbourne Valley, Surrey 1560–1760', *Southern History*, v. 6 (1984) pp. 79–83.
123 Arthur Raistrick, *Industrial Archaeology* (1973) pp. 256–7 C J. D. Marshall and John K. Walton, *The Lake Counties from 1830 to the mid-twentieth century* (1981) pp. 27–8. Paul N. Wilson, 'The Gunpowder Mills of Westmorland and Furness', *Transactions of the Newcomen Society*, v. 36 (1963–4) pp. 47–65.
124 *Oxford Dictionary of National Biography* (2004).
125 J. G. Moher, 'John Torr Foulds (1742–1815): millwright and engineer', *Transactions of the Newcomen Society*, v. 58 (1986–7) pp. 59–73.
126 P. F. Brandon, 'Land, Technology and Water Management in the Tillingbourne Valley, Surrey 1560–1760', *Southern History*, v. 6 (1984) pp. 75–6.
127 D. C. Watts, 'Water Power in the Industrial Revolution', *Transactions of the Cumberland and Westmoreland Antiquarian and Archaeological Society*, new series v. 67 (1966) pp. 199–205.
128 Stanley D. Chapman, 'The Cost of Power in the Industrial Revolution in Britain: the case of the textile industry', *Midland History*, v. 1, no. 2 (Autumn 1971) p. 10.
129 E. J. T. Collins, ed., *The Agrarian History of England and Wales*, v. 7, 1850–1914 (2000) pp. 386–7.
130 Hills, *Power in the Industrial Revolution*, pp. 49, 67–9. Arthur Raistrick, *Industrial Archaeology* (1973) pp. 149–50. *Oxford Dictionary of National Biography* (2004), entry on Sir Richard Arkwright. R. S. Fitton, *The Arkwrights, Spinners of Fortune* (1989) pp. 27–8.
131 Hills, *Power in the Industrial Revolution*, pp. 69–70. R. S. Fitton and A. P. Wadsworth, *The Strutts and the Arkrights 1758–1830: A study of the early factory system* (1958) pp. 64–5.
132 Hills, *Power in the Industrial Revolution*, pp. 69–70, 91. R. S. Fitton, *The Arkwrights, Spinners of Fortune* (1989) pp. 81, 190. S. D. Chapman, 'The Arkwright Mills: Colquhoun's census of 1788 and archaeological evidence', *Industrial Archaeological Review*, v. 6 (1981–2) pp. 5–27.
133 Hills, *Power in the Industrial Revolution*, pp. 106–10. Mary B. Rose, *The Gregs of Quarry Bank Mill: The rise and decline of a family firm 1750–1914* (1986) pp. 24–5, 38.

134 H. D. Gribbon, *The History of Water Power in Ulster* (1969) pp. 110ff.
135 William Pole, ed., *The Life of Sir William Fairbairn, Bart, partly written by himself* (1877, reprinted with introduction by A. E. Musson, 1970) pp. 121–3, 313–14. William Fairbairn, *Mills and Millwork*, I, pp. 88–9, 130–1.
136 Hills, *Power in the Industrial Revolution* (1970) p. 211.
137 W. B. Crump and G. Ghorbal, *History of the Huddersfield Woollen Industry* (1935) pp. 68–9.
138 D. T. Jenkins, 'Early Factory Development in the West Riding of Yorkshire 1770–1800', *Textile History and Economic History: Essays in honour of Miss Julia de Lacy Mann*, ed. N. B. Harte and K. G. Ponting (1973) pp. 253–63. John Goodchild, 'The Ossett Mill Company', *Textile History*, v. 1 (1968–70) pp. 46–61.
139 D. T. Jenkins, 'Early Factory Development in the West Riding of Yorkshire 1770–1800', *Textile History and Economic History: Essays in honour of Miss Julia de Lacy Mann*, ed. N. B. Harte and K. G. Ponting (1973) pp. 263–5, 279–80.
140 Frederick J. Glover, 'The Rise of the Heavy Woollen Trade of the West Riding of Yorkshire in the Nineteenth Century', *Business History*, v. 4 (1961) p. 2.
141 Jennifer Tann, *Gloucestershire Woollen Mills* (1967) pp. 51–4. Jennifer Tann, 'The Employment of Power in the West-of-England Wool Textile Industry', *Textile History and Economic History: Essays in honour of Miss Julia de Lacy Mann*, ed. N. B. Harte and K. G. Ponting (1973) pp. 198–200. Marilyn Palmer and Peter Neaverson, *The Textile Industry of South-west England: A social archaeology* (2005) pp. 71–5.
142 D. T. Jenkins and K. G. Ponting, *The British Wool Textile Industry 1770–1914* (1982) pp. 52–6.
143 D. T. Jenkins and K. G. Ponting, *The British Wool Textile Industry 1770–1914* (1982) pp. 55–6.
144 G. E. Mingay, ed., *The Agrarian History of England and Wales*, v. 6, 1750–1850 (1989) p. 386. E. J. T. Collins, *The Economy of Upland Britain, 1750–1950: an illustrated review* (1978) p. 20.
145 W. A. McCutcheon, 'Water Power in the North of Ireland', *Transactions of the Newcomen Society*, v. 39 (1966–7) p. 78.
146 H. D. Gribbon, *The History of Water Power in Ulster* (1969) pp. 80–90. Enid Gauldie, 'Water-powered Beetling Machines', *Transactions of the Newcomen Society*, v. 39 (1966–7) pp. 125–8.
147 McCutcheon, *North of Ireland*, pp. 78–83. H. D. Gribbon, *The History of Water Power in Ulster* (1969) pp. 90–102.
148 McCutcheon, *North of Ireland*, pp. 73–7. H. D. Gribbon, *The History of Water Power in Ulster* (1969) pp. 102–9.
149 Hills, *Power in the Industrial Revolution*, p. 93.
150 Hills, *Power in the Industrial Revolution*, p. 193.
151 Hills, *Power in the Industrial Revolution*, p. 97.
152 Barrie Trinder, *The Industrial Revolution in Shropshire* (1981 edn) pp. 18, 21.
153 Barrie Trinder, *The Industrial Revolution in Shropshire* (1981 edn) pp. 39–40, 46–7, 99.
154 H. D. Gribbon, *The History of Water Power in Ulster* (1969) pp. 76–9.
155 simt.co.uk. T. S. Ashton, *An Economic History of England: The 18th century* (1955) p. 94.
156 R. A. Pelham, 'The Water Power Crisis in Birmingham in the Eighteenth Century', *University of Birmingham Historical Journal*, v. 9 (1961) pp. 80–3.
157 Rex Wailes, 'Water-driven Mills for Grinding Stone', *Transactions of the Newcomen Society*, v. 39 (1966–7) pp. 95–119. Peter Lord, *Portrait of the River Trent* (1968) p. 46.
158 Martin Watts, 'Sawmills: a slice through time', *Mill News*, no. 115 (April 2008) pp. 25–8.
159 Martin Watts, p. 98.
160 Arthur Raistrick and Bernard Jennings, *A History of Lead Mining in the Pennines* (1965) pp. 140, 152, 211–12, 215–17, 219–23, 229, 234, 236, 238–9.
161 N. V. Allen, *The Waters of Exmoor* (1978) pp. 19–20.
162 E. J. T. Collins, ed., *The Agrarian History of England and Wales*, v. 7, 1850–1914 (2000) p. 1133. J. D. Marshall and John K. Walton, *The Lake Counties from 1830 to the mid-twentieth century* (1981) pp. 12, 24–7.
163 H. W. Strong, *Industries of North Devon* (1889) pp. 67–74.
164 Charles Hadfield, *The Canal Age* (1968) p. 55.
165 Martin Watts p. 75.
166 Martin Watts pp. 75–6.
167 *Railway* magazine, September 1978, p. 426.
168 Hills, *Power in the Industrial Revolution*, p. 113. Frank Nixon, *The Industrial Archaeology of Derbyshire* (1969) p. 99.
169 Reynolds, *Stronger than a Hundred Men*, pp. 205–12.
170 Norman A. F. Smith, 'The Origins of the Water Turbine and the Invention of its Name', *History of Technology*, v. 2 (1977) p. 221.
171 Hills, *Power in the Industrial Revolution*, p. 99. W. Fairbairn, *Mills and Millwork*, I, p. 113. P. N. Wilson, 'The Waterwheels of John Smeaton', *Transactions of the Newcomen Society*, v. 30 (1955–57) pp. 25–48.
172 Reynolds, *Stronger than a Hundred Men*, pp. 278–86.
173 For example, William Donaldson, *Principles of Construction and Efficiency of Water-wheels* (1876) pp. 19–21.
174 W. Fairbairn, *Mills and Millwork*, I, pp. 88–9, 113.
175 W. Fairbairn, *Mills and Millwork*, I. pp. 123–4.
176 William Pole, ed., *The Life of Sir William Fairbairn, Bart, partly written by himself* (1877, reprinted with

Notes

introduction by A. E. Musson, 1970) pp. 188–9. W. Fairbairn, *Mills and Millwork*, I. pp. 133–7.
177 *The Engineer*, v. 22 (Jul–Dec 1866) pp. 81–2.
178 W. Fairbairn, *Mills and Millwork*, I. pp. 151–3.
179 *Implement and Machinery Review*, v. 27 (1901–2) p. 430.
180 Martin Watts, pp. 75, 100. H. D. Gribbon, *The History of Water Power in Ulster* (1969) pp. 134–5. N. V. Allen, *The Waters of Exmoor* (1978) p. 18. Joseph Glynn, *Rudimentary Treatise on the Power of Water as applied to drive Flour Mills and to give Motion to Turbines and other Hydrostatic Engines* (1885 edn) pp. 83–4.
181 H. D. Gribbon, *The History of Water Power in Ulster* (1969) pp. 21, 25. Rhodes Boyson, *The Ashworth Cotton Enterprise: The rise and fall of a family firm 1818–1880* (1970) pp. 19–22.
182 H. D. Gribbon, *The History of Water Power in Ulster* (1969) p. 19.
183 Hills, *Power in the Industrial Revolution*, pp. 110–13. H. R. Johnson and A. W. Skempton, 'William Strutt's Cotton Mills 1793–1812', *Transactions of the Newcomen Society*, v. 30 (1955–7) p. 190.
184 Glynn, *Rudimentary Treatise on the Power of Water*, p. 84. Giles and Goodall pp. 128–9.
185 Frank Nixon, *The Industrial Archaeology of Derbyshire* (1969) p. 99. Glynn, *Rudimentary Treatise on the Power of Water*, pp. 91–2.
186 Tunzelmann, *Steam Power*, p. 128.
187 W. Fairbairn, *Mills and Millwork*, I, pp. 125–6.
188 W. Fairbairn, *Mills and Millwork*, I, pp. 129–31.
189 Enid Gauldie, *The Scottish Country Miller*, p. 122.
190 Leslie Syson p. 54. P. N. Wilson, 'The Waterwheels of John Smeaton', *Transactions of the Newcomen Society*, v. 30 (1955–57) p. 33.
191 P. N. Wilson, 'The Waterwheels of John Smeaton', pp. 25–48.
192 Hills, *Power in the Industrial Revolution*, pp. 107–12.
193 R. S. Fitton and A. P. Wadsworth, *The Strutts and the Arkwrights 1758–1830: A study of the early factory system* (1958) pp. 210–11.
194 J. Harold Armfield, 'The Craft of Mill Gearing', *Industrial Archaeology*, v. 9 (1972) pp. 296–300.
195 Reynolds, *Stronger than a Hundred Men*, pp. 290–2.
196 Martin Watts, p. 77.
197 Quoted in Orbell, *The Corn Milling Industry*, p. 287.
198 Hills, *Power in the Industrial Revolution*, p. 113. William Pole, ed., *The Life of Sir William Fairbairn, Bart, partly written by himself* (1877, reprinted with introduction by A. E. Musson, 1970) pp. 121–2.
199 W. Fairbairn, *Mills and Millwork*, v. 1, p. 129.
200 Martin Watts pp. 78–80.
201 Harrison, *Eight Centuries of Milling*, pp. 69–72.
202 Alain Belmont, 'Millstone Production and Trade in Europe as an Indicator of Changing Agricultural Practice and Human Diet', John Broad, ed., *A Common Agricultural Heritage? Revising French and British rural divergence* (2009) pp. 189–91.
203 Harrison, *Eight Centuries of Milling*, pp. 99–100.
204 Michael Stratton and Barrie Trinder, *Industrial England* (1997) p. 29.
205 G. N. von Tunzelmann, *Steam Power and British Industrialization to 1860 (1978)* p. 128. Rhodes Boyson, *The Ashworth Cotton Enterprise: The rise and fall of a family firm 1818–1880* (1970) pp. 61–2, 94, 96.
206 Adam Menuge, 'The Cotton Mills of the Derbyshire Derwent and its Tributaries', *Industrial Archaeology Review*, v. 16 (1993) p. 49. Anthony Calladine and Jean Fricker, *East Cheshire Textile Mills* (1993) p. 34.
207 Reynolds, *Stronger than a Hundred Men*, p. 268. G. N. von Tunzelmann, *Steam Power and British Industrialization to 1860* (1978) p. 137, quoting PP 1846, VI, Q.7588. and p. 138.
208 Hills, *Power in the Industrial Revolution*, pp. 104–5.
209 John Farey, *General View of the Agriculture of Derbyshire* (1806) v. 2 p. 492. Paul N. Wilson, 'British Industrial Waterwheels', *Transactions of the International Symposium on Molinology*, v. 3 (1973) p. 27. Arthur Raistrick, *Industrial Archaeology* (1972) pp. 251–3.
210 Tunzelmann, *Steam Power*, p. 172. W. Fairbairn, *Mills and Millwork*, I, pp.131–2. Glynn, *Rudimentary Treatise on the Power of Water*, pp. 30–3. Trevor Turpin, *Dam* (2008) pp. 34–5.
211 Harrison, *Eight Centuries of Milling*, pp. 98–9.
212 Giles and Goodall p. 132. Trevor Turpin, *Dam* (2008) pp. 143–5.
213 Martin Watts, pp. 109–10.
214 H. D. Gribbon, *The History of Water Power in Ulster* (1969) p. 28.
215 Martin Watts p.110.
216 H. D. Gribbon, *The History of Water Power in Ulster* (1969) pp. 29–30.
217 P. N. Wilson, 'Gilkes's 1853–1975: 122 years of water turbine and pump manufacture', *Transactions of the Newcomen Society*, v. 47 (1974–6) pp. 73–84.
218 Paul N. Wilson, 'Early Water Turbines in the United Kingdom', *Transactions of the Newcomen Society*, v. 31 (1957–59) pp, 234–5.
219 Joseph Addison and Rex Wailes, 'Dorset Watermills', *Transactions of the Newcomen Society*, v. 35 (1962–3) pp. 211–12. M. D. Freeman, 'Armfield Turbines in Hampshire', *Industrial Archaeology*, v. 8 (1971) pp. 373–80.
220 Martin Watts, pp. 113–14. Chris Page, 'C. L. Hett of Brigg: hydraulic engineer', in Ken Redmore, ed., *Ploughs, Chaff Cutters and Steam Engines: Lincolnshire's agricultural implement makers* (2007) pp. 85–94.
221 Joseph Addison, 'Second Addendum to Dorset Watermills', *Transactions of the Newcomen Society*, v. 41 (1968–9) pp. 149–50. J. R. Gray, 'An Agricultural

222 Farm Estate in Berkshire', *Industrial Archaeology*, v. 8 (1971) pp. 177–8, 180.
222 Arthur Raistrick, *Industrial Archaeology* (1973) p. 117.
223 Arthur Raistrick and Bernard Jennings, *A History of Lead Mining in the Pennines* (1965) pp. 219–23. Ian McNeill, *Hydraulic Power* (1972).
224 John Blake Ltd, Hydram catalogue, c.1950.
225 For fuller details about mill buildings of the Industrial Revolution *see* Colum Giles and Ian H. Goodall, *Yorkshire Textile Mills* (1992) and Anthony Calladine and Jean Fricker, *East Cheshire Textile Mills* (1993).
226 Giles and Goodall, *Yorkshire Textile Mills*, pp. 126–7.
227 Edgar Jones, *Industrial Architecture in Britain 1750–1939* (1985) pp. 15–35.
228 Peter Wenham, *Watermills* (1989) p. 79. Edgar Jones, *Industrial Architecture in Britain 1750–1939* (1985) p. 110.
229 B. R. Mitchell and Phyllis Deane, *Abstract of British Historical Statistics* (1962) pp. 5–9.
230 Orbell, *The Corn Milling Industry*, pp. 273–4. G. E. Mingay, ed., *The Agrarian History of England and Wales*, v. 6, 1750–1850 (1989) p. 395.
231 Watts, *Water and Wind Power*, p. 102. William Coles Finch, *Watermills and Windmills: A historical survey of their rise and fall as portrayed by those of Kent* (1933) p. 37.
232 Watts, *Water and Wind Power*, p. 92.
233 Mingay, *Agrarian History*, v. 6, p. 404.
234 Orbell, *The Corn Milling Industry*, p. 278.
235 Orbell, *The Corn Milling Industry*, pp. 42, 241, 249. Dave Gregory and Michael Jones, *Norwell Mills* (Norwell Heritage Booklet 3, 2009) pp. 13–15.
236 Owen Ashmore, *The Industrial Archaeology of North West England* (1982) p. 34.
237 Mingay, *Agrarian History*, v. 6, pp. 386–7. R. Thurston Hopkins, *Old Watermills and Windmills* (nd, c.1920) p. 160. Derek Stidder and Colin Smith, *Watermills of Sussex*, vol. 1, East Sussex (1997) pp. 55, 70–1, 147–8.
238 Orbell, *The Corn Milling Industry*, p. 279. Charles Hadfield, *The Canal Age* (1968) p. 55. wiltonwindmill.co.uk.
239 *The Story of Haxted Mill*.
240 Orbell, *The Corn Milling Industry*, pp. 279–81.
241 Orbell, *The Corn Milling Industry*, pp. 285–6.
242 L. Caroe, 'Urban Change in East Anglia in the Nineteenth Century', PhD thesis, University of Cambridge (1966) p. 107.
243 *Berkshire Mercury*, 12 April 1824, advertisement of mill to let.
244 Hervey Benham, *Some Essex Water Mills* (1976, 2nd edn 1983) p. 41.
245 John Shaw, *Water Power in Scotland 1550–1870* (1984) pp. 161–2. Charles Vancouver, *General View of the Agriculture of the County of Hampshire* (1813) pp. 109–10. N. V. Allen, *The Waters of Exmoor* (1978) p. 22.
246 John Holt, *General View of the Agriculture of the County of Lancaster* (1795) pp. 34–5. Arthur Young, *General View of the Agriculture of the County of Lincolnshire* (1813) p. 95. Martin Watts, *Water and Wind Power* (2002) p. 103. Joseph Addison and Rex Wailes, 'Dorset Watermills', *Transactions of the Newcomen Society*, v. 35 (1962–3) pp. 203. N. V. Allen, *The Waters of Exmoor* (1978) pp. 16–18.
247 Roy Brigden, *Victorian Farms* (1986) pp. 48–50, 62. Martin Watts, *Water and Wind Power* (2002) pp. 103–4.
248 N. V. Allen, *The Waters of Exmoor* (1978) p. 22.
249 P. N. Wilson, 'Gilkes's 1853–1975: 122 years of water turbine and pump manufacture', *Transactions of the Newcomen Society*, v. 47 (1974–6) p. 74. Joseph Addison, 'Second Addendum to Dorset Watermills', *Transactions of the Newcomen Society*, v. 41 (1968–9) pp. 146–7.
250 W. G. Rimmer, *Marshalls of Leeds, Flax Spinners 1788–1886* (1960) pp. 26–7, 34–5.
251 A. Ure, *Philosophy of Manufactures* (1835) p. 9, quoted by Hills, *Power in the Industrial Revolution*, p. 95.
252 Tunzelmann, *Steam Power*, p. 170. Harrison, *Eight Centuries of Milling*, p. 15. Giles and Goodall, pp. 129–30.
253 Derek Gregory, *Regional Transformation and Industrial Revolution: A geography of the Yorkshire woollen industry* (1982) pp. 68–70, 208. Tunzelmann, *Steam Power*, p. 173.
254 Mary B. Rose, *The Gregs of Quarry Bank Mill: The rise and decline of a family firm 1750–1914* (1986) pp. 42–3. Adam Menuge, 'The Cotton Mills of the Derbyshire Derwent and its Tributaries', *Industrial Archaeology Review*, v. 16 (1993) p. 41.
255 Hills, *Power in the Industrial Revolution*, pp. 105–6.
256 W. Fairbairn, *Treatise on Mills and Millwork* (1861) I, pp. 88–90.
257 D. T. Jenkins and K. G. Ponting, *The British Wool Textile Industry 1770–1914* (1982) p. 120.
258 Hills, *Power in the Industrial Revolution*, pp. 128, 129, 131. Mary B. Rose, *The Gregs of Quarry Bank Mill: The rise and decline of a family firm 1750–1914* (1986) pp. 42–3, 143–4.
259 Tann, 'The Employment of Power', pp. 200–1. Anthony Calladine and Jean Fricker, *East Cheshire Textile Mills* (1993) p. 131.
260 Stanley D. Chapman, 'The Cost of Power in the Industrial Revolution in Britain: the case of the textile industry', *Midland History*, v. 1, no. 2 (Autumn 1971) pp. 4–11, 16–19.
261 Tunzelmann, *Steam Power*, pp. 53–62. Tann, 'The Employment of Power', pp. 215–20. Alistair G.

Thomson, *The Paper Industry in Scotland 1590–1861* (1974) p. 55.
262 B. R. Mitchell and Phyllis Deane, *Abstract of British Historical Statistics* (1962) pp. 480–3, tables 7 and 8.
263 SC on Manufacturers' Employment, PP 1830, v. X, p. 223.
264 Tunzelmann, *Steam Power*, pp. 62–7. Tann, 'The Employment of Power', p. 207.
265 John W. Kanefsky, 'Motive Power in British Industry and the Accuracy of the 1870 Factory Return', *Economic History Review*, 2nd ser. v. 32 (1979) p. 369.
266 Mary B. Rose, *The Gregs of Quarry Bank Mill: The rise and decline of a family firm 1750–1914* (1986) pp. 42, 143–4. Tunzelmann, *Steam Power*, pp. 135–6, 161–3.
267 Mary B. Rose, *The Gregs of Quarry Bank Mill: The rise and decline of a family firm 1750–1914* (1986) pp. 42, 143–4. W. G. Rimmer, *Marshalls of Leeds, Flax Spinners 1788–1886* (1960) pp. 133–4. Derek Gregory, *Regional Transformation and Industrial Revolution: A geography of the Yorkshire woollen industry* (1982) p. 203.
268 Hills, *Power in the Industrial Revolution*, pp. 101–3, 136–7, 161, citing references in the Boulton & Watt records. Adam Menuge, 'The Cotton Mills of the Derbyshire Derwent and its Tributaries', *Industrial Archaeology Review*, v. 16 (1993) p. 52. Reynolds, *Stronger than a Hundred Men*, pp. 322–5.
269 Rhodes Boyson, *The Ashworth Cotton Enterprise: The rise and fall of a family firm 1818–1880* (1970) pp. 15, 21–2.
270 Martin Watts, p. 99–100.
271 Boyson, *The Ashworth Cotton Enterprise*, pp. 61–2.
272 Boyson, *The Ashworth Cotton Enterprise*, pp. 61–2. G. N. von Tunzelmann, *Steam Power and British Industrialization to 1860* (1978) p. 33. A. E. Musson, 'Industrial Motive Power in the United Kingdom 1800–1870', *Economic History Review*, 2nd ser., v. 29 (1976) pp. 431–3.
273 John W. Kanefsky, 'Motive Power in British Industry and the Accuracy of the 1870 Factory Return', *Economic History Review*, 2nd ser. v. 32 (1979) pp. 360–75.
274 Giles and Goodall, *Yorkshire Textile Mills*, p. 125. Derek Gregory, *Regional Transformation and Industrial Revolution: A geography of the Yorkshire woollen industry* (1982) pp. 201–3. D. T. Jenkins and K. G. Ponting, *The British Wool Textile Industry 1770–1914* (1982) pp. 119–21.
275 Jenkins and Ponting, *The British Wool Textile Industry*, pp. 54–5. John Shaw, *Water Power in Scotland 1550–1870* (1984) pp. 285–6.
276 Jenkins and Ponting, *The British Wool Textile Industry*, pp. 119–20.
277 Jenkins and Ponting, *The British Wool Textile Industry*, pp. 121–2. H. D. Gribbon, *The History of Water Power in Ulster* (1969) pp. 61–8.
278 Jenkins and Ponting, *The British Wool Textile Industry*, pp. 122–4.
279 J. D. Marshall and John K. Walton, *The Lake Counties from 1830 to the mid-twentieth century* (1981) pp. 23–4.
280 E. J. T. Collins, ed., *The Agrarian History of England and Wales*, v. 7, 1850–1914 (2000) p. 1110.
281 H. D. Gribbon, *The History of Water Power in Ulster* (1969) pp. 61–8, 187–8, 191.
282 Marilyn Palmer and Peter Neaverson, *The Textile Industry of South-west England: A social archaeology* (2005) pp. 122–3.
283 R. C. Riley, 'Henry Cort at Funtley, Hampshire', *Industrial Archaeology*, v. 8 (1971) pp. 69–76. John F. Potter, 'Iron Working in the Vicinity of Weybridge, Surrey', *Industrial Archaeological Review*, v. 6 (1981–2) pp. 211–23.
284 A. H. Shorter, *Paper Making in the British Isles* (1971) pp. 110–12, 158. Alistair G. Thomson, *The Paper Industry in Scotland 1590–1861* (1974) pp. 55n, 159.
285 E. J. T. Collins, ed., *The Agrarian History of England and Wales*, v. 7, 1850–1914 (2000) p. 22.
286 John W. Kanefsky, 'Motive Power in British Industry and the Accuracy of the 1870 Factory Return', *Economic History Review*, 2nd ser. v. 32 (1979) p. 368. Marilyn Palmer and Peter Neaverson, *The Textile Industry of South-west England: A social archaeology* (2005) pp. 123ff. Martin Watts, p. 102.
287 John W. Kanefsky, 'Motive Power in British Industry and the Accuracy of the 1870 Factory Return', *Economic History Review*, 2nd ser. v. 32 (1979) pp. 368–9.
288 John W. Kanefsky, 'Motive Power in British Industry and the Accuracy of the 1870 Factory Return', *Economic History Review*, 2nd ser. v. 32 (1979) p. 370.
289 Thomas Crump, *Age of Steam*, p.61.
290 Orbell, *The Corn Milling Industry*, pp. 169–70.
291 Orbell, *The Corn Milling Industry*, pp. 170–1.
292 Quoted in Orbell, *The Corn Milling Industry*, pp. 171–2.
293 Orbell, *The Corn Milling Industry*, pp. 63–4, 173. G. E. Mingay, ed., *The Agrarian History of England and Wales*, v. 6, 1750–1850 (1989) p. 395.
294 E. Marriage & Son Ltd, *The Annals of One Hundred Years of Flour Milling* (1940).
295 M. Diane Freeman, 'Assessing Potential Milling Capacity in Hampshire c1750–1914', *Industrial Archaeology Review*, v. 1 (1976–7) p. 54.
296 E. J. T. Collins, ed., *The Agrarian History of England and Wales*, v. 7, 1850–1914 (2000) p. 1063. H. D. Gribbon, *The History of Water Power in Ulster* (1969) pp. 44–9.

297 M. D. Freeman, 'A History of Corn Milling …', thesis, pp. 83–4.

298 H. W. Strong, *Industries of North Devon* (1889) pp. 195–205. *Implement and Machinery Review*, v. 35 (1909–10) pp. 212–13.

299 R. G. Burnett, *Through the Mill* (1945) pp. 26ff. John W. Kanefsky, 'Motive Power in British Industry and the Accuracy of the 1870 Factory Return', *Economic History Review*, 2nd ser. v. 32 (1979) p. 372. M. Diane Freeman, 'Assessing Potential Milling Capacity in Hampshire c1750–1914', *Industrial Archaeology Review*, v. 1 (1976–7) pp. 47–62. H. D. Gribbon, *The History of Water Power in Ulster* (1969) pp. 44, 49.

300 M. D. Freeman, 'A History of Corn Milling with special reference to south central and south eastern England', PhD thesis, University of Reading, 1976, p. 136.

301 McCutcheon, *North of Ireland*, p. 73. Harrison, *Eight Centuries of Milling*, p. 189.

302 Sue Owen, *et al.*, *Rivers and the British Landscape* (2005) pp. 177–8. Hadrian Cook, 'A Tale of Two Catchments: water management and quality in the Wandle and Tillingbourne 1600–1990', *Southern History*, v. 30 (2008) pp. 94, 98–9. Nicholas Goddard, 'The Suburbanization of the English Landscape: environmental conflict in Victorian Croydon', P. S. Barnwell and Marilyn Palmer, ed., *Post-Medieval Landscapes*, v. 3 (2007) pp. 120–1. John Sheail, 'Underground Water Abstraction: indirect effects of urbanization in the countryside', *Journal of Historical Geography*, v. 8 (1982) pp. 395–408.

303 Rex Wailes, 'Water-driven Mills for Grinding Stone', *Transactions of the Newcomen Society*, v. 39 (1966–7) p. 100. Derek Stidder and Colin Smith, *Watermills of Sussex*, vol. 1, East Sussex (1997) pp. 69–70.

304 Ministry of Housing and Local Government, 'Tax on Watermills: the effects of the Water Resources Act 1963', *Transactions of the International Symposium on Molinology*, v. 2 (1969) pp. 145–60. Mike Davies-Shiel, *Watermills of Cumbria* (1978) pp. 22–3.

305 Joseph Addison, 'Second Addendum to Dorset Watermills', *Transactions of the Newcomen Society*, v. 41 (1968–9) p. 141. David Llewewllyn Davies, *Watermill: Life story of a Welsh cornmill* (1997) pp. 35–6. *Daily Telegraph*, 18 June 1983.

306 Sue Owen, *et al.*, *Rivers and the British Landscape* (2005) p. 1.

307 Stuart Downward and Kenneth Skinner, 'Working Rivers: the geomorphological legacy of English freshwater mills', *Area*, v. 37 (2005) pp. 138–47. John K. Harrison, *Eight Centuries of Milling in North East Yorkshire* (2nd edn 2008) p. 17. Des O'Sullivan, 'Too Precious to Lose: the unique river environment around Taplow Mill', *Hitcham and Taplow Society Newsletter*, no. 91 (Spring 2009) pp. 10–11.

308 Henry Cleere and David Crossley, *The Iron Industry of the Weald* (1985) pp. 221–42.

309 Kenneth C. Reid, *Water Mills and the Landscape* (nd) pp. 3–4. Marilyn Palmer and Peter Neaverson, *The Textile Industry of South-west England: A social archaeology* (2005) pp. 32–3.

310 H. C. Darby, ed., *A New Historical Geography of England before 1600* (1976) p. 245, citing Carus-Wilson, *Expansion of Exeter* (1963) p. 22.

311 R. Thurston Hopkins, *Old Watermills and Windmills* (nd, c.1920) pp. 113–21. Hadrian Cook, 'A Tale of Two Catchments: water management and quality in the Wandle and Tillingbourne 1600–1990', *Southern History*, v. 30 (2008) p. 96.

312 Hills, *Power in the Industrial Revolution*, p. 94.

313 Hills, *Power in the Industrial Revolution*, p. 105.

314 John Farey, *General View of the Agriculture of Derbyshire* (1813) v. 2, p. 398.

315 Quoted in R. S. Fitton and A. P. Wadsworth, *The Strutts and the Arkwrights 1758–1830: A study of the early factory system* (1958) p. 223.

316 R. S. Fitton, *The Arkwrights, Spinners of Fortune* (1989) pp. 81–4.

317 J. D. Marshall and John K. Walton, *The Lake Counties from 1830 to the mid-twentieth century* (1981) pp. 27–28.

318 Arthur Raistrick, *Industrial Archaeology* (1972) pp. 246–53.

319 Pyrs Gruffudd, '"Uncivil Engineering": nature, nationalism and hydro-electrics in north Wales', in Denis Cosgrove and Geoff Petts, ed., *Water, Engineering and Landscape* (1990) pp. 159–73.

320 Museum of English Rural Life, Council for the Protection of Rural England archives, SR CPRE C1/62/79.

321 David Blackbourn, *The Conquest of Nature: Water, landscape and the making of modern Germany* (2006) pp. 216–19.

322 Allan Gifford, *Derbyshire Watermills: Corn mills* (1999) p. 126.

323 Henrietta Heald, 'Lord of the Crags', *The Spectator*, 23 June 2007, pp. 59–60.

324 G. Woodward, 'Hydro-electricity in North Wales 1880–1948', *Transactions of the Newcomen Society*, v. 69 (1997–8) p. 205. N. V. Allen, *The Waters of Exmoor* (1978) p. 21.

325 D. G. Tucker, 'Hydro-electricity for Public Supply in Britain 1881–1894', *Industrial Archaeology Review*, v.1 (1977) pp. 126–63. Martin Watts pp. 114–15.

326 McCutcheon, *North of Ireland*, p. 86.

327 Martin Watts, p. 114.

328 McCutcheon, *North of Ireland*, p. 88. H. D. Gribbon, *The History of Water Power in Ulster* (1969) pp. 147–55.

329 G. Woodward, 'Hydro-electricity in North Wales', *Transactions of the Newcomen Society*, v. 69 (1997–8) pp. 205–7.
330 Woodward, 'Hydro-electricity in North Wales', p. 208.
331 Woodward, 'Hydro-electricity in North Wales', pp. 209–12. E. J. T. Collins, ed., *The Agrarian History of England and Wales*, v. 7, 1850–1914 (2000) p. 14.
332 The preceding paragraphs draw upon Woodward, 'Hydro-electricity in North Wales', pp. 218–26.
333 Woodward, 'Hydro-electricity in North Wales', pp. 216–18.
334 E. J. T. Collins, *The Economy of Upland Britain, 1750–1950: An illustrated review* (1978) pp. 25–6.
335 Leslie Hannah, *Electricity before Nationalisation* (1979) pp. 119–21, 129–31.
336 Leslie Hannah, *Engineers Managers and Politicians: The first fifteen years of nationalised electricity supply in Britain* (1982) pp. 149–50.
337 Leslie Hannah, *Engineers Managers and Politicians*, pp.152–3. Godfrey Boyle, ed., *Renewable Energy: Power for a sustainable future* (2004) pp. 148–50.
338 H. D. Gribbon, *The History of Water Power in Ulster* (1969) pp. 140–43.
339 H. D. Gribbon, *The History of Water Power in Ulster* (1969) pp. 145–6.
340 Geoffrey Binnie, 'The Evolution of British Dams', *Transactions of the Newcomen Society*, v. 47 (1974–6) pp. 208–9, 220–1.
341 R. Clare, ed., *Tidal Power: Trends and development* (1992) pp. 27–168. McCutcheon, *North of Ireland*, p. 88.
342 Severn Tidal Power Group, *Tidal Power from the Severn* (1986).
343 *The Times* online, 27 January 2009, *Daily Telegraph* online, 27 January 2009, *The Guardian* online, 26 January 2009.
344 Richard Bennett and John Elton, *History of Corn Milling*, v. 2 (1899), p. 202. William Coles Finch, *Watermills and Windmills: A historical survey of their rise and fall as portrayed by those of Kent* (1933) pp. 35–42.
345 Martin Watts, *Water and Wind Power* (2002) p. 119.
346 William Coles Finch, *Watermills and Windmills: A historical survey of their rise and fall as portrayed by those of Kent* (1933) p. 38.
347 Raistrick, pp. 284, 289.
348 *The Guardian*, 28 July 1999.
349 Scottish & Southern Electricity website: scottish-southern.co.uk.
350 Paul N. Wilson, 'The Gunpowder Mills of Westmorland and Furness', *Transactions of the Newcomen Society*, v. 36 (1963–4) p. 59.
351 J. M. Beskine, 'Generating electricity on the farm', *Farm Mechanization*, v. 1 (1946–7) pp. 248–51.
352 Martin Watts p. 118.
353 Commission for the European Communities, Community Energy Technology Projects in the Sector of Hydro-electric Energy (1993).
354 *The Times*, 20 July 1991.
355 Martin Watts p. 115.
356 *Daily Telegraph*, 19 August 2008, 18 April 2009.
357 *Daily Telegraph, Weekend*, 3 January 2009.
358 Godfrey Boyle, ed., *Renewable Energy: Power for a sustainable future* (2004) p. 176.
359 *Sunday Telegraph*, 2 November 2008, p. L9. Websites of Water Power Enterprises, Torrs Hydro, and Esk Energy.
360 G. Muller and K. Krauppert, 'Old Watermills – Britain's new source of energy?', Proceedings of the Institution of Civil Engineers, *Civil Engineering*, v. 150 (November 2002) pp. 178–86.
361 Godfrey Boyle, ed., *Renewable Energy: Power for a sustainable future* (2004) p. 175.
362 *Daily Telegraph* online, 29 November, 18 December 2008.
363 reuk.co.uk/Tidal-Turbines.htm Accessed 28 January 2009. Godfrey Boyle, ed., *Renewable Energy: Power for a sustainable future* (2004) pp. 230–4.
364 Godfrey Boyle, ed., *Renewable Energy: Power for a sustainable future* (2004) pp. 298–9, 301–2. Commission for the European Communities, Wave Energy Research and Development: proceedings of a workshop held at Cork 1 and 2 October 1992 (1993) pp. 12–13. Leon Freris and David Infield, *Renewable Energy in Power Systems* (2008) p. 48.
365 'Milestone for Wave Energy Plans', bbc.co.uk 16 March 2010.

Glossary

Beetling mill. In the linen industry, a mill in which cloth was subject to pounding to produce a sheen.

Bleach green. Term used especially in Ireland for a mill that housed the beetling and bleaching processes for linen manufacture.

Bloomery. A forge in which iron blooms (bars) are produced by smelting and hammering.

Blowing mill. A works for producing iron by smelting, in which bellows, often powered by waterwheel, direct air to the furnace.

Breast-shot wheel. A vertical waterwheel for which the water enters high on the near side, turning the wheel anti-clockwise, by impulse or gravity. Depending on the point at which the water hits the wheel, this type is variously known as a low breast-shot or high breast-shot (*see* also 'Pitch-back').

Bucket. The container on overshot or breast-shot waterwheels into which water was dropped to turn it.

Cam. A projection on a wheel or shaft, which trips another part of machinery to set it in motion.

Clasp arm. A method of construction for wheels in which arms fixed across the wheel form a clamp around the hub. This is in contrast to having a set of spokes radiating from the centre (compass arm). Clasp-arm wheels were used for both waterwheels and gear wheels.

Compass-arm wheel. A wheel with shafts that radiate from the centre more or less along the lines of the main compass points.

Crazing mill. The term in the tin industry for a mill that grinds metal ore.

Crown wheel. A horizontal gear, usually at the top of an upright shaft, which engages with pinions to drive horizontal layshafts.

Demesne. The part of a manorial estate managed directly.

Edge-runner stone. A stone set to revolve on its edge, to crush grains, ore and stones.

Fall. The effective operational head of water, the distance it descended in doing the work.

Felloe. A curved section of the rim of a wheel.

Float (or float board). The paddle on an undershot wheel.

Fulling mill. A mill for beating woollen cloth to cleanse it and bind its fibres together. The beating was by heavy hammers, or fulling stocks.

Goit (goyt). A term used especially in Yorkshire for the millstream or leat.

Governor. A device for regulating the speed at which waterwheels turn. It was also applied to the windmill and the steam engine, and in flour mills used to determine the gap between millstones.

Gravity wheel. A waterwheel driven by the weight of water on its blades or buckets.

Head. The difference in height between the water coming on to the wheel and the outflow.

Headrace. The mill channel upstream of the wheel, bringing the water from the mainstream.

Horizontal wheel. A waterwheel that turns in the horizontal plane.

Horsepower. A unit of power introduced by James Watt to demonstrate the capacity of his steam engines, defined as the power required to lift 33,000lb 1 foot in a minute. It was used as a measure of power for most engines and machines, including waterwheels.

Impulse wheel. A wheel driven by the force of water on its blades. This could apply to undershot waterwheels, and some types of turbine.

Incorporating mill. A mill in which the ingredients of gunpowder are pulverized and ground together under edge-runner stones.

Lade. An alternative term for the millstream, or leat, mainly used in Scotland.

Lantern. A gear made of two discs with rods between them that engage with the teeth of an opposed gear. This was a principal drive gear in ancient and medieval mills. Also called a 'trundle'.

Launder. The artificial channel that brings water overhead to an overshot waterwheel.

Layshaft. A main drive shaft that runs in the horizontal plane. Gearing from this shaft takes the drive to individual machines.

Leat. The artificial channel constructed to bring water from the river to the mill, also known as the 'millrace'. The part above the channel above the wheel is the 'headrace'; below is the 'tailrace'.

Glossary

Millrace. The feeder channel to the mill, also known as the 'leat'.

Millstones. The stones between which grain is ground to flour.

Millwright. The craftsman who built and maintained the mill.

Multure. The payment in grain made to the miller by the tenant of a manorial estate.

Noria. A water-lifting device consisting of a wheel turned by the propulsion of a stream on paddles, enabling buckets to scoop up water and carry it to the higher level. Developed in the classical Near East, this was probably the earliest water-powered device.

Overshot wheel. Vertical waterwheel fed from above, the water falling into a bucket, to turn the wheel by gravity.

Penstock. Sluice gate to regulate the admission of water to the wheel. Also used to describe the artificial chute that directs water to the wheel. For overshot wheels this is often called a 'launder'.

Pentrough. Artificial channel bringing water to the wheel.

Pitch-back wheel. A waterwheel with water entering at the top of the circle, turning it anti-clockwise, the opposite direction to the overshot wheel. It is thus a very high breast-shot wheel, and might also be known by this term.

Pitwheel. The main drive gear in most watermills. It is usually parallel to the waterwheel, and partially sunk into the pit, from which it derives its name.

Quern. A hand-powered tool for grinding grain into flour.

Race. The mill channel, or leat.

Reaction turbine. A water turbine in which the water is forced into passages to turn the blades.

Rim gear. A primary drive gear from the waterwheel. The gear engages first with cogs around the rim of the waterwheel and second with another gear to transmit the drive forwards.

Runner. The moving stone of a pair of millstones.

Rynd. An iron bar with 'T' section head that connects the vertical drive shaft to the millstone.

Scouring. The term for finishing the point of a needle.

Scribbling. One of the processes for drawing out the fibres of wool in preparation for spinning. Scribbling mills housed the machines for this process.

Scutching. Breaking down fibres of flax preparatory to making it into linen.

Sliding hatch. A penstock that operates by depressing the gate to allow water to flow over the top of the hatch on to the wheel.

Slitting mill. Works in the iron industry for cutting pig iron into bars.

Soke mill. A manorial mill to which the tenants of the estate were obliged to bring the grain to be ground.

Spurwheel. A central gear that enables more than one drive to be taken from it. It was especially used in corn mills to allow more than one pair of stones to be driven by the one waterwheel.

Stamps. Vertical bars of wood or iron lifted by cams to act as beaters, to crush ores and other raw materials.

Stocks. The heavy hammers used in fulling.

Suspension wheel. A waterwheel of iron, in which solid spokes are replaced by tension rods to tie the hub to the rim.

Tailrace. The mill channel downstream of the waterwheel, which returns the water to the mainstream.

Thirlage. Scottish feudal term for the system by which tenants of an estate were bound to bring their corn to the lord's mill for grinding.

Tide mill. Watermill operated by the flow of the tide.

Trundle. Alternative term for lantern gear.

Tucking mill. Term used in the West Country for a fulling mill.

Undershot wheel. Vertical waterwheel driven by the water passing underneath it to turn it by impulse.

Vertical wheel. A waterwheel that revolves in the vertical plane.

Wallower. A horizontal gear engaging with the pitwheel to transmit the drive to an upright shaft.

Water turbine. Water engine that involves a wheel revolving at high speed, almost always a horizontal wheel.

Watt. A basic unit of power, being the amount of energy expended in one second. It was introduced for the measurement of electrical power, but also applied to waterwheels. Kilowatt (KW): 1,000 watts (or 746hp); megawatt (MW): million watts; gigawatt (GW): thousand million watts.

Weir. A low dam across a stream, to divert water to the watermill.

Bibliography

ARCHIVAL SOURCES

John Blake Ltd, catalogue for Hydrams, *c.* 1950
Council for the Protection of Rural England Collection, Museum of English Rural Life
Fitz Waterwheel Co, Hanover, Pennsylvania, Catalogue no. 60, Water Power on the Farm, 1923
The John Vince Collection, Museum of English Rural Life
Williamson Brothers, catalogues

NEWSPAPERS AND JOURNALS

Berkshire Mercury
The Guardian
Daily Telegraph
The Engineer
Implement and Machinery Review
The Miller
Milling
Mill News
The Spectator
The Times
Vintage Spirit

DIRECTORIES

Kelly's Directory, various editions
Pigot's Directory, 1830

BOOKS AND ARTICLES

Joseph Addison and Rex Wailes, 'Dorset Watermills', *Transactions of the Newcomen Society*, v. 35 (1962–3) pp. 193–216
Joseph Addison, 'Addendum to Dorset Watermills', *Transactions of the Newcomen Society*, v. 36 (1963–4) pp. 175–81
Joseph Addison, 'Second Addendum to Dorset Watermills', *Transactions of the Newcomen Society*, v. 41 (1968–9) pp. 139–62
N. V. Allen, *The Waters of Exmoor* (1978)
Peter Allen, 'How Paper-making Came to Colthrop', *Berkshire Old and New*, no. 4 (1987), pp. 8–16
K. J. Allison, *East Riding Watermills* (1970)
J. Harold Armfield, 'The Craft of Mill Gearing', *Industrial Archaeology*, v. 9 (1972) pp. 296–300
Owen Ashmore, *The Industrial Archaeology of Lancashire* (1969)
Owen Ashmore, *The Industrial Archaeology of North West England* (1982)
T. S. Ashton, *An Economic History of England: the 18th century* (1955)
G. G. Astill, *A Medieval Industrial Complex and its Landscape: The metalworking watermills and workshops of Bordesley Abbey* (1993)
Brian G. Awty, 'The Cumbrian Bloomery Forge in the Seventeenth Century and Forge Equipment in the Charcoal Iron Industry', *Transactions of the Newcomen Society*, v. 51 (1979–80) pp. 25–41
George Basalla, *The Evolution of Technology* (1988)
Alain Belmont, 'Millstone Production and Trade in Europe as an Indicator of Changing Agricultural Practice and Human Diet', in John Broad, ed., *A Common Agricultural Heritage? Revising French and British rural divergence* (2009) pp. 171–92
Hervey Benham, *Some Essex Watermills* (1976, 2nd edn 1983)
C. E. Bennett, 'The Watermills of Kent, East of the Medway', *Industrial Archaeology Review*, v. 1 (1977) pp. 205–35
Richard Bennett and John Elton, *History of Corn-Milling* (4 vols, 1898–1904)
Maurice Beresford and John Hurst, *Wharram Percy Deserted Medieval Village* (1990)
J. M. Beskine, 'Generating electricity on the farm', *Farm Mechanization*, v. 1 (1946–7) pp. 248–51
Geoffrey Binnie, 'The Evolution of British Dams', *Transactions of the Newcomen Society*, v. 47 (1974–6) pp. 207–24
David Blackbourn, *The Conquest of Nature: Water, landscape and the making of modern Germany* (2006)

Bibliography

Edmund Blunden, 'Country Childhood', in Simon Nowell-Smith, ed., *Edwardian England* (1964)

Guy Blythman, *Watermills and Windmills of Middlesex* (1996)

J. L. Bolton, *The Medieval English Economy 1150–1500* (1980)

C. J. Bond, 'The Reconstruction of the Medieval Landscape: The estates of Abingdon Abbey', *Landscape History*, v. 1 (1979) pp. 59–75

C. T. G. Boucher, 'Broadstone Mill, Corvedale', *Transactions of the Newcomen Society*, v. 36 (1963–4) pp. 159–63

Godfrey Boyle, ed., *Renewable Energy: Power for a sustainable future* (2004)

Rhodes Boyson, *The Ashworth Cotton Enterprise: The rise and fall of a family firm 1818–1880* (1970)

P. F. Brandon, 'Land, Technology and Water Management in the Tillingbourne Valley, Surrey 1560–1760', *Southern History*, v. 6 (1984) pp. 75–103

Roy Brigden, *Victorian Farms* (1986)

R. G. Burnett, *Through the Mill* (1945)

Anthony Calladine, 'Lombe's Mill: An exercise in reconstruction', *Industrial Archaeology Review* v. 16 (1993) pp. 82–99

Anthony Calladine and Jean Fricker, *East Cheshire Textile Mills* (1993)

L. Caroe, 'Urban Change in East Anglia in the Nineteenth Century', PhD thesis, University of Cambridge (1966)

E. M. Carus-Wilson, 'An Industrial Revolution of the Thirteenth Century', *Economic History Review*, v. 7 (1941) pp. 39–60

Stanley D. Chapman, 'The Cost of Power in the Industrial Revolution in Britain: The case of the textile industry', *Midland History*, v. 1, no. 2 (Autumn 1971) pp. 1–23

S. D. Chapman, 'The Arkwright Mills: Colquhoun's census of 1788 and archaeological evidence', *Industrial Archaeological Review*, v. 6 (1981–2) pp. 5–27

Carlo M. Cipolla, *Before the Industrial Revolution: European society and economy, 1000–1700* (1976)

R. Clare, ed., *Tidal Power: Trends and development* (1992)

Gillian Clark, *Down by the River: The Thames and Kennet in Reading* (2009)

Kenneth Clark, *Civilisation* (1969)

Henry Cleere and David Crossley, *The Iron Industry of the Weald* (1985)

D. C. Coleman, *The British Paper Industry, 1495–1860* (1958)

E. J. T. Collins, *The Economy of Upland Britain, 1750–1950: An illustrated review* (1978)

Commission for the European Communities, 'Wave Energy Research and Development: Proceedings of a workshop held at Cork 1 and 2 October 1992' (1993)

Commission for the European Communities, 'Community Energy Technology Projects in the Sector of Hydro-Electric Energy' (1993)

Hadrian Cook, 'A Tale of Two Catchments: Water management and quality in the Wandle and Tillingbourne 1600–1990', *Southern History*, v. 30 (2008) pp. 78–103

D. W. Crossley, ed., *Medieval Industry* (1981)

D. W. Crossley, 'Medieval Iron Smelting', in D. W. Crossley, ed., *Medieval Industry* (1981) pp. 29–41

Thomas Crump, *A Brief History of the Age of Steam* (2007)

W. B. Crump and G. Ghorbal, *History of the Huddersfield Woollen Industry* (1935)

E. Cecil Curwen, 'The Problem of Early Watermills', *Antiquity*, v. 18 (1944) pp. 130–46

H. C. Darby, ed., *A New Historical Geography of England before 1600* (1976)

H. C. Darby, *Domesday England* (1986 edn)

David Llewewllyn Davies, *Watermill: Life story of a Welsh corn-mill* (1997)

Mike Davies-Shiel, *Watermills of Cumbria* (1978)

Joan Day and R. F. Tylecote, *The Industrial Revolution in Metals* (1991)

Daniel Defoe, *A Tour through the Whole Island of Great Britain* (1724–6, 1974 edn)

Daniel Defoe, *The Complete English Tradesman* (1745)

Department of Energy, Central Electricity Generating Board and Severn Tidal Power Group, 'The Severn Barrage Project: general report' (1989)

H. S. L. Dewar, 'The Windmills, Watermills and Horse Mills of Dorset', *Proceedings of the Dorset Natural History & Archaeological Society*, v. 82 (1960) pp. 109–32

D. Dilworth, *The Tame Mills of Staffordshire* (1976)

William Donaldson, *Principles of Construction and Efficiency of Water-Wheels* (1876)

Stuart Downward and Kenneth Skinner, 'Working Rivers: the geomorphological legacy of English freshwatermills', *Area*, v. 37 (2005) pp. 138–47

Christopher Dyer, *Making a Living in the Middle of Ages: The people of Britain 850–1520* (2003)

George Eliot, *The Mill on the Floss* (1860)

Monica Ellis, ed., *Water and Wind Mills in Hampshire and the Isle of Wight* (1978)

W. English, *The Textile Industry* (1969)

Nesta Evans, *The East Anglian Linen Industry: Rural industry and local economy 1500–1850* (1985)

Robert L. Evans, *Fuelling our Future: An introduction to sustainable energy* (2007)

William Fairbairn, *Treatise on Mills and Millwork* (Part I 1861, Part II 1863)

Rosamond Faith, 'The "Great Rumour" of 1377 and Peasant Ideology', in *The English Rising of 1381*, ed., R. H. Hilton and T. H. Aston (1984) pp. 43–73

Keith A. Falconer, 'Fireproof Mills: the widening perspective', *Industrial Archaeology Review*, v. 16 (1993) pp. 11–26

John Farey, *General View of the Agriculture of Derbyshire* (1813)

William Coles Finch, *Watermills and Windmills: A historical survey of their rise and fall as portrayed by those of Kent* (1933)

R. S. Fitton and A. P. Wadsworth, *The Strutts and the Arkwrights 1758–1830: A study of the early factory system* (1958)

R. S. Fitton, *The Arkwrights, Spinners of Fortune* (1989)

John Fitzherbert, *Surveyinge* (1539)

R. J. Forbes, *Studies in Ancient Technology* (1955)

M. D. Freeman, 'Armfield Turbines in Hampshire', *Industrial Archaeology*, v. 8 (1971) pp. 373–80

M. D. Freeman, 'Funtley Iron Mill, Fareham, Hants', *Industrial Archaeology*, v. 8 (1971) pp. 63–8

M. D. Freeman, 'A History of Corn-milling with special reference to south central and south eastern England', PhD thesis, University of Reading, 1976

M. Diane Freeman, 'Assessing Potential Milling Capacity in Hampshire c1750–1914', *Industrial Archaeology Review*, v. 1 (1976–7) pp. 47–62

Leon Freris and David Infield, *Renewable Energy in Power Systems* (2008)

M. J. Fuller, *The Watermills of the East Malling and Wateringbury Streams* (1973)

E. M. Gardner, *The Three Mills: Tide mills part three* (1957)

Enid Gauldie, 'Water-powered Beetling Machines', *Transactions of the Newcomen Society*, v. 39 (1966–7) pp. 125–8

Enid Gauldie, *The Scottish Country Miller 1700–1900: A history of water-powered meal milling in Scotland* (1981)

Allan Gifford, *Derbyshire Watermills: Corn mills* (1999)

Colum Giles and Ian H. Goodall, *Yorkshire Textile Mills* (1992)

Bertrand Gille, *The Renaissance Engineers* (English edn 1966)

Jean Gimpel, *The Medieval Machine* (1976)

Frederick J. Glover, 'The Rise of the Heavy Woollen Trade of the West Riding of Yorkshire in the Nineteenth Century', *Business History*, v. 4 (1961) pp. 1–22

Joseph Glynn, *Rudimentary Treatise on the Power of Water as applied to drive Flour Mills and to give Motion to Turbines and other Hydrostatic Engines* (1885 edn)

Nicholas Goddard, 'The Suburbanization of the English Landscape: Environmental conflict in Victorian Croydon', P. S. Barnwell and Marilyn Palmer, ed., *Post-Medieval Landscapes*, v. 3 (2007) pp. 119–34

John Goodchild, 'The Ossett Mill Company', *Textile History*, v. 1 (1968–70) pp. 46–61

Alan H. Graham, 'The Old Malthouse, Abbotsbury, Dorset: the medieval watermill of the Benedictine Abbey', *Proceedings of the Dorset Natural History and Archaeological Society*, v. 108 (1986) pp. 103–25

J. R. Gray, 'An Agricultural Farm Estate in Berkshire', *Industrial Archaeology*, v. 8 (1971) pp. 175–85

Nan Greatrex, 'The Robinson Enterprise at Papplewick, Nottinghamshire', *Industrial Archaeology Review*, v. 9 (1986–7) pp. 37–56, 119–39

Kevin Greene, 'Technological Innovation and Economic Progress in the Ancient World: M. I. Finlay reconsidered', *Economic History Review*, v. 53 (2000) pp. 29–59

T. A. P. Greeves, 'The Archaeological Potential of the Devon Tin Industry', in D. W. Crossley, ed., *Medieval Industry* (1981) pp. 85–95

Dave Gregory and Michael Jones, *Norwell Mills* (Norwell Heritage Booklet 3, 2009)

Derek Gregory, *Regional Transformation and Industrial Revolution: A geography of the Yorkshire woollen industry* (1982)

H. D. Gribbon, *The History of Water Power in Ulster* (1969)

Pyrs Gruffudd, '"Uncivil Engineering": Nature, nationalism and hydro-electrics in north Wales', in Denis Cosgrove and Geoff Petts, ed., *Water, Engineering and Landscape* (1990) pp. 159–73

Charles Hadfield, *The Canal Age* (1971 edn)

A. Rupert Hall and N. C. Russell, 'What about the Fulling Mill?' *History of Technology*, v. 6 (1981) pp. 113–19

Leslie Hannah, *Electricity before Nationalisation* (1979)

Leslie Hannah, *Engineers Managers and Politicians: The first fifteen years of nationalised electricity supply in Britain* (1982)

Helen Harris, *The Industrial Archaeology of the Peak District* (1971)

John K. Harrison, *Eight Centuries of Milling in North East Yorkshire* (2nd edn 2008)

Jeff Hawksley, *The Power and the Glory of Waterwheels* (2008)

Richard Hayman, 'Artists' Impressions of Aberdulais Mill', *Industrial Archaeology Review*, v. 9 (1986–7) pp. 155–66

Donald Hill, *A History of Engineering in Classical and Medieval Times* (1984)

R. L. Hills, *Power in the Industrial Revolution* (1970)

R. L. Hills, *Papermaking in Britain 1488–1988* (1988)

Margaret T. Hodgen, 'Domesday Watermills', *Antiquity*, v. 13 (1939) pp. 261–79

John Holt, *General View of the Agriculture of the County of Lancaster* (1795)

Richard Holt, *The Mills of Medieval England* (1988)

Richard Holt, 'Medieval England's water-related technologies', *Working with Water in Medieval Europe: Technology and resource use*, ed. Paolo Squatriti (2000), pp. 51–100

R. Thurston Hopkins, *Old Watermills and Windmills* (nd, c1920)

K. Howarth, 'Weir Construction on the River Irwell at Warth, Radcliffe', *Industrial Archaeology*, v. 8 (1971) pp. 240–6

Thomas P. Hughes, *Networks of Power: electrification in western society 1880–1930* (1983)

D. T. Jenkins, 'Early Factory Development in the West Riding of Yorkshire 1770–1800', *Textile History and Economic History: Essays in honour of Miss Julia de Lacy Mann*, ed. N. B. Harte and K. G. Ponting (1973) pp. 247–80

D. T. Jenkins and K. G. Ponting, *The British Wool Textile Industry 1770–1914* (1982)

J. G. Jenkins, *The Welsh Woollen Industry* (1969)

J. G. Jenkins, ed., *The Wool Textile Industry in Great Britain* (1972)

H. R. Johnson and A. W. Skempton, 'William Strutt's Cotton Mills 1793–1812', *Transactions of the Newcomen Society*, v. 30 (1955–57) pp. 179–205

W. Branch Johnson, *Industrial Archaeology of Hertfordshire* (1970)

David H. Jones, 'Manx Watermills', *Journal of the Manx Museum*, v. 7 (1966) pp. 11–16

David Jones, 'The Water-powered Cornmills of England, Wales and the Isle of Man: a preliminary account of their development', *Transactions of the International Symposium on Molinology*, v. 2 (1969) pp. 303–54

D. H. Jones, 'Belbroughton: a water-powered edge tool district', *Transactions of the International Symposium on Molinology*, v. 8 (1993) pp. 17–26

Edgar Jones, *Industrial Architecture in Britain 1750–1939* (1985)

John W. Kanefsky, 'Motive Power in British Industry and the Accuracy of the 1870 Factory Return', *Economic History Review*, 2nd ser. v. 32 (1979) pp. 360–75

J. G. Landels, *Engineering in the Ancient World* (1978)

John Landers, *The Field and the Forge: Population, production and power in the pre-industrial west* (2003)

John Langdon, *Mills in the Medieval Economy* (2004)

Serge Leliavsky, *Dams* (1981)

Serge Leliavsky, *Weirs* (1981)

M. J. T. Lewis, *Millstone and Hammer* (1997)

M. J. T. Lewis, W. N. Slatcher and P. N. Jarvis, 'Flashlocks in English Waterways: a survey', *Industrial Archaeology*, v. 6 (1969) pp. 209–53

Peter Lord, *Portrait of the River Trent* (1968)

David Luckhurst, *Monastic Watermills* (nd)

W. A. McCutcheon, 'Water Power in the North of Ireland', *Transactions of the Newcomen Society*, v. 39 (1966–7) pp. 67–94

Adrian T. McDonald and David Kay, *Water Resources: Issues and strategies* (1988)

Thomas McErlean and Norman Crothers, *Harnessing the Tides: The early medieval tide mills at Nendrum Monastery, Strangford Lough* (2007)

Ian McNeill, *Hydraulic Power* (1972)

J. Kenneth Major, 'Berkshire Watermills', *Berkshire Archaeological Journal*, v. 61 (1963–4) pp. 83–92

E. Marriage & Son Ltd, *The Annals of One Hundred Years of Flour-milling* (1940)

J. D. Marshall and John K. Walton, *The Lake Counties from 1830 to the Mid-Twentieth Century* (1981)

Adam Menuge, 'The Cotton Mills of the Derbyshire Derwent and its Tributaries', *Industrial Archaeology Review*, v. 16 (1993) pp. 38–61

W. T. Miller, *The Watermills of Sheffield* (4th edn 1949)

G. E. Mingay, ed., *The Agrarian History of England and Wales*, v. 6, 1750–1850 (1989)

Ministry of Housing and Local Government, 'Tax on Watermills: the effects of the Water Resources Act 1963', *Transactions of the International Symposium on Molinology*, v. 2 (1969) pp. 145–60

B. R. Mitchell and Phyllis Deane, *Abstract of British Historical Statistics* (1962)

J. G. Moher, 'John Torr Foulds (1742–1815): Millwright and engineer', *Transactions of the Newcomen Society*, v. 58 (1986–7) pp. 59–73

Joel Mokyr, ed., *The Oxford Encyclopaedia of Economic History* (2003)

L. A. Moritz, *Grain-Mills and Flour in Classical Antiquity* (1958)

G. Muller and K. Krauppert, 'Old Watermills – Britain's new source of energy?', Proceedings of the Institution of Civil Engineers, *Civil Engineering*, v. 150 (November 2002) pp. 178–86

Lewis Mumford, *Technics and Civilization* (1934)

A. E. Musson and Eric Robinson, *Science and Technology in the Industrial Revolution* (1969)

A. E. Musson, 'Industrial Motive Power in the United Kingdom 1800–1870', *EcHR*, 2nd ser., v. 29 (1976) pp. 415–39

A. E. Musson, *The Growth of British Industry* (1978)

Frank Nixon, *The Industrial Archaeology of Derbyshire* (1969)

W. L. Norman, 'The Wakefield Soke Mills to 1853', *Industrial Archaeology*, v. 7 (1970) pp. 176–83

J. P. Oleson, *Greek and Roman Mechanical Water-lifting Devices: The history of a technology* (1984)

M. J. Orbell, 'The Corn-milling Industry in the Industrial Revolution', PhD thesis, University of Nottingham (1977)

Des O' Sullivan, 'Too Precious to Lose: the unique river environment around Taplow Mill', *Hitcham and Taplow Society Newsletter*, no. 91 (Spring 2009) pp. 10–11

Sue Owen, et al., *Rivers and the British Landscape* (2005)

Oxford Dictionary of National Biography (2004)

Chris Page, 'C. L. Hett of Brigg: hydraulic engineer', in Ken Redmore, ed., *Ploughs, Chaff Cutters and Steam Engines: Lincolnshire's agricultural implement makers* (2007) pp. 85–94

Marilyn Palmer and Peter Neaverson, *The Textile Industry of South-west England: A social archaeology* (2005)

R. A. Pelham, 'The Water Power Crisis in Birmingham in the Eighteenth Century', *University of Birmingham Historical Journal*, v. 9 (1961) pp. 62–91

William Pole, ed., *The Life of Sir William Fairbairn, Bart, partly written by himself* (1877, reprinted with introduction by A. E. Musson, 1970)

K. G. Ponting, *Wool and Water* (1975)

John F. Potter, 'Iron Working in the Vicinity of Weybridge, Surrey', *Industrial Archaeological Review*, v. 6 (1981–2) pp. 211–23

Timothy E. Powell, 'The Disappearance of Horizontal Watermills from Medieval Ireland', *Transactions of the Newcomen Society*, v. 66 (1994–5) pp. 219–24

Uvedale Price, *Essays in the Picturesque* (1810)

P. A. Rahtz, 'Medieval Milling', in D. W. Crossley, ed., *Medieval Industry* (1981) pp. 1–15

Arthur Raistrick, *Industrial Archaeology* (1972)

Arthur Raistrick and Bernard Jennings, *A History of Lead Mining in the Pennines* (1965)

G. D. Ramsey, *The Wiltshire Woollen Industry during the Seventeenth Century* (1943)

Abraham Rees, *Cyclopaedia or Universal Dictionary of Arts, Sciences and Literature* (1802–19); selected items republished as *Rees's Manufacturing Industry*, ed. N. Cossons (1972)

Kenneth C. Reid, *Watermills and the Landscape* (nd)

John Reynolds, *Windmills and Watermills* (1970)

J. Reynolds, C. Burrell and D. Bignell, *Durngate Mill, Winchester* (nd)

Terry Reynolds, *Stronger than a Hundred Men: A history of the vertical waterwheel* (1983)

P. S. Richards, 'The Holywell Textile Mills, Flintshire', *Industrial Archaeology*, v. 6 (1969) pp. 28–51

R. C. Riley, 'Henry Cort at Funtley, Hampshire', *Industrial Archaeology*, v. 8 (1971) pp. 69–76

W. G. Rimmer, *Marshalls of Leeds, Flax Spinners 1788–1886* (1960)

John G. Rollins, *The Needle Mills* (1970)

Mary B. Rose, *The Gregs of Quarry Bank Mill: The rise and decline of a family firm 1750–1914* (1986)

Rural Industries Bureau, *Water Power* (nd, 1930s)

Colin Rynne, 'The Introduction of the Vertical Watermill into Ireland: some recent archaeological evidence', *Medieval Archaeology*, v. 33 (1989) pp. 21–31

H. R. Schubert, *History of the British Iron and Steel Industry from c450BC to AD1775* (1957)

Select Committee on Manufacturers' Employment, Parliamentary Papers, 1830, v. X, pp. 221–33

Severn Tidal Power Group, *Tidal Power from the Severn* (1986)

John Shaw, *Water Power in Scotland 1550–1870* (1984)

John Sheail, 'Underground Water Abstraction: indirect effects of urbanization in the countryside', *Journal of Historical Geography*, v. 8 (1982) pp. 395–408

R. Shorland-Ball, 'Worsborough Corn Mill, South Yorkshire', *Industrial Archaeology Review*, v. 2 (1978) pp. 240–64

A. H. Shorter, *Paper-Making in the British Isles* (1971)

Charles Singer, E. J. Holmyard, A. R. Hall and Trevor I. Williams, ed., *A History of Technology* (1958)

Josef Sisitka, 'Floating mills in London: an historical survey', unpublished BA dissertation, University of Reading (1992)

Josef Sisitka, 'Floating mills in London: an historical survey', *Industrial Archaeology Review*, v. 19 (1997) pp. 21–30

C. P. Skilton, *British Windmills and Watermills* (1947)

K. Smith, *Water in Britain* (1972)

Norman Smith, *A History of Dams* (1971)

Norman Smith, *Man and Water* (1976)

Norman A. F. Smith, 'The Origins of the Water Turbine and the Invention of its Name', *History of Technology*, v. 2 (1977) pp. 215–59

N. A. F. Smith, 'The Origins of Water Power: a problem of evidence and expectations', *Transactions of the Newcomen Society*, v. 55 (1983–4) pp. 67–84

R. J. Spain, 'Romano-British Watermills', *Archaeologia Cantiana*, v. 100 (1984) pp. 101–28

R. J. Spain, 'The Second-century Romano-British Corn Mill at Ickham, Kent', *History of Technology*, v. 9 (1984) pp. 143–80

John M. Steane, *The Archaeology of Medieval England and Wales* (1984)

Derek Stidder and Colin Smith, *Watermills of Sussex*, vol. 1, East Sussex (1997)

Derek Stidder and Colin Smith, *Watermills of Sussex*, vol. 2, West Sussex (2001)

The Story of Haxted Mill

Arthur Stowers, 'Observations on the History of Water Power', *Transactions of the Newcomen Society*, v. 30 (1955–7) pp. 239–56

Michael Stratton and Barrie Trinder, *Industrial England* (1997)

Michael Stratton and Barrie Trinder, *Twentieth-Century Industrial Archaeology* (2000)

H. W. Strong, *Industries of North Devon* (1889)

Leslie Syson, *British Watermills* (1965)

Jennifer Tann, 'Some Problems of Water Power: A study of mill siting', *Transactions of the Bristol and Gloucestershire Archaeological Society*, v. 84 (1965) pp. 53–77

Jennifer Tann, *Gloucestershire Woollen Mills* (1967)

Jennifer Tann, 'The Employment of Power in the West-of-England Wool Textile Industry', *Textile History and Economic History: Essays in honour of Miss Julia de Lacy Mann*, ed. N. B. Harte and K. G. Ponting (1973) pp. 198–224

Jennifer Tann and Glyn Jones, 'Technology and Transformation: the diffusion of the roller mill in the British flour-milling industry 1870–1907', *Technology and Culture*, v. 37 (1996) pp. 36–69

Fred S. Thacker, *Kennet Country* (1932)

Joan Thirsk, *Alternative Agriculture* (1997)

Alistair G. Thomson, *The Paper Industry in Scotland 1590–1861* (1974)

Martin Tillmans, *Bridge Hall Mills: Three centuries of paper and cellulose film manufacture* (1978)

Charles Tomlinson, *Cyclopaedia of useful arts, mechanical and chemical, manufactures, mining and engineering* (1851)

Barrie Trinder, *The Industrial Revolution in Shropshire* (1981 edn)

D. G. Tucker, 'Millstones, Quarries and Millstone-makers', *Post-medieval Archaeology*, vol. 11 (1977) pp. 1–21

D. G. Tucker, 'Hydro-electricity for Public Supply in Britain 1881–1894', *Industrial Archaeology Review*, v.1 (1977) pp. 126–63

D. G. Tucker, 'Millstone Making in the Peak District of Derbyshire: the quarries and the technology', *Industrial Archaeology Review*, v.8 (1985) pp. 42–58

D. G. Tucker, 'Millstone Making in England', *Industrial Archaeology Review*, v.9 (1986–7) pp. 167–82

Bibliography

G. N. von Tunzelmann, *Steam Power and British Industrialization to 1860* (1978)

Trevor Turpin, *Dam* (2008)

H. R. Vallentine, *Water in the Service of Man* (1967)

C. Vancouver, *General View of the Agriculture of Hampshire* (1813)

Victoria County History, *Hampshire*, v. 5 (1912)

Rex Wailes, Tide Mills, parts 1 and 2 (nd). Pamphlet reprint of article from *Transactions of the Newcomen Society*, v. 19 (1938–9)

Rex Wailes, 'Suffolk Watermills', *Transactions of the Newcomen Society*, v. 37 (1964–5) pp. 99–116

Rex Wailes, 'Water-driven Mills for Grinding Stone', *Transactions of the Newcomen Society*, v. 39 (1966–7) pp. 95–119

D. C. Watts, 'Water-power in the Industrial Revolution', *Transactions of the Cumberland and Westmoreland Antiquarian and Archaeological Society*, new series v. 67 (1966) pp. 199–205

Martin Watts, *Water and Wind Power* (2002)

Martin Watts, 'Sawmills: a slice through time', *Mill News*, no. 115 (April 2008) pp. 25–8

Peter Wenham, *Watermills* (1989)

Lynn White Jr, *Medieval Technology and Social Change* (1962)

Orjan Wikander, 'Archaeological Evidence for Early Watermills – an interim report', *History of Technology*, v. 10 (1985) pp. 151–80

James E. Williams, 'Howsham Watermill', *York Georgian Society Annual Report*, 1966–7, pp. 12–15

Kenneth Williamson, 'Horizontal Watermills of the Faeroe Islands', *Antiquity*, v. 20 (1946) pp. 83–91

Paul N. Wilson, *Watermills: An introduction* (nd)

P. N. Wilson, 'The Waterwheels of John Smeaton', *Transactions of the Newcomen Society*, v. 30 (1955–57) pp. 25–48

Paul N. Wilson, 'Early Water Turbines in the United Kingdom', *Transactions of the Newcomen Society*, v. 31 (1957–9) pp. 219–41

Paul N. Wilson, *Watermills with Horizontal Wheels* (1960)

Paul N. Wilson, 'The Gunpowder Mills of Westmorland and Furness', *Transactions of the Newcomen Society*, v. 36 (1963–4) pp. 47–65

Paul N. Wilson, 'British Industrial Waterwheels', *Transactions of the International Symposium on Molinology*, v. 3 (1973) pp. 17–31

P. N. Wilson, 'Gilkes's 1853–1975: 122 years of water turbine and pump manufacture', *Transactions of the Newcomen Society*, v. 47 (1974–6) pp. 73–84

Wiltshire County Council and Wilton Windmill Society, *Wilton Windmill* (1979)

G. Woodward, 'Hydro-electricity in North Wales 1880–1948', *Transactions of the Newcomen Society*, v. 69 (1997–8) pp. 205–35

Jim Woodward-Nutt, ed., *Mills Open* (2007 edn)

Neil Wright, *Lincolnshire Towns and Industry 1700–1914* (1982)

Margaret Yates, 'Watermills in the Local Economy of a Late-Medieval Manor in Berkshire', in T. Thornton, ed., *Social Attitudes and Political Structures in the Fifteenth Century* (2000)

Margaret Yates, *Town and Country in Western Berkshire c1327–c1600: Social and economic change* (2007)

Andrew Yeats and Eric Parks, 'The Eco-renovation of Gibson Mill', *Building* (Summer 2006) pp. 12–18

Arthur Young, *General View of the Agriculture of the County of Lincolnshire* (1813)

C. Young, 'Excavations at Ickham', *Archaeologia Cantiana*, v. 91 (1975) pp. 190–1

WEBSITES

Aquamarine Power Ltd, aquamarinepower.com
BBC, bbc.co.uk
The British Hydropower Association, british-hydro.org
Daily Telegraph, telegraph.co.uk
Department of Energy and Climate Change, decc.gov.uk
Dorset Against Rural Turbines, dartdorset.org
Environment Agency, environment-agency.gov.uk
Esk Energy, Ruswarp project, northyorkmoors-npa.gov.uk/eskenergy
First Hydro Company, fhc.co.uk
Gazetteer for Scotland, geo.ed.ac.uk/scotgaz
The Guardian, guardian.co.uk
Hampshire Mills Group, hampshiremills.org
Norfolk Mills, norfolkmills.co.uk
Ramblers Holidays, Lemsford Mill, ramblersholidays.co.uk/lemsford_mill.aspx
REUK, reuk.co.uk/Worlds-First-Open-Sea-Tidal-Turbine.htm
Royal Gunpowder Mills, Waltham Abbey, royalgunpowdermills.com
Scottish and Southern Electricity, scottish-southern.co.uk
Sheffield Industrial Museums Trust, simt.co.uk
The Times Online, timesonline.co.uk
The Traditional Cornmillers Guild, tcmg.org.uk
Wandle Industrial Museum, wandle.org
Water Power Enterprises, h2ope.org.uk
Wilton windmill, wiltonwindmill.co.uk
Yorkshire Post online, yorkshirepost.co.uk

Index

Abbeydale 84, 178
Abbotsbury, Dorset 46, 49, 49
Aberdare 98
Aberdeenshire 139, 148, 151, 152
Abingdon 55, 56, 128
Abingdon Abbey 19, 41, 44
Aire River 79
Albion flour mills 153–4
Allenheads lead mines 121
Alne River 67
Alresford 21, 34
Aluminium smelting 171, 172, 174
Alvingham, Lincolnshire 132
Anker, River 19
Antipater 16, 17
Archimedean screw 15, 15, 182, 183
Archimedes 15
Arkwright, Richard 14, 44, 74–5, 76, 82, 88, 107, 112, 114, 123, 124, 139, 140, 141, 142, 143, 166, 167, 180
Arkwright, Richard II 139
Armfield, manufacturers, Ringwood 119, 120, 120
Armstrong, Sir William 119, 121, 169, 170
Arrow River 67, 84
Ashford, Derbyshire 86
Ashworth family, mill-owners 96, 113, 146
Ashworth, Henry 147
Avon & Stour Catchment Board 160
Avon River 67
Aymestrey, Herefordshire 159
Ayrshire 71, 91, 108

Bacheldore Mill 180
Backbarrow, Furness 74
Bakewell, Derbyshire 44, 76, 107, 139, 143
Balmoral 182
Bann River 94, 100, 115, 174
Bark crushing 47, 57, 58, 60
Barker wheel 116, 118
Barton on Windrush 59
Beeleigh, Essex 129
Beetling 64, 81
Beighton, Henry 108–9, 112
Belbroughton, Worcestershire 84
Belfast 77, 108, 118, 119, 150, 155
Belper, Derbyshire 11, 11, 76, 82, 88, 97, 97, 103, 104, 106, 108, 114, 116, 143, 145, 166, 166, 167, 179, 181
Bennet, John, of Newbury 60
Bennett, Richard 17, 177
Berkshire 18, 21, 22, 62, 64, 69, 94, 120
Berwickshire 152
Bessbrook & Newry Tramway 171
Bidston, Cheshire 50, 52
Birdham, Sussex 53
Birmingham 56, 67, 73, 84, 113, 142, 154, 182
Blade mill 67
Blaenau Ffestiniog 171
Blast furnace 37, 65, 83, 104
Bleaching, linen 81, 02, 113, 150, 174
Blockley, Gloucestershire 21
Bloomery 65
Boatmen 42–3, 114
Bobbin making 71, 87
Bollin River 76, 90, 165
Bolting machine 111, 112
Bolton 82, 96, 113, 146, 147
Bonsall Brook 74, 75, 114
Bordesley Abbey 65
Botley, Hampshire 131
Boulton & Watt 32, 85, 121, 138, 141, 142–4, 146, 151, 153
Boulton, Matthew 32, 84–5
Bradford 24, 79, 80, 140, 144
Bradford upon Avon 149
Bradiford, Devon 87
Brailes, Worcestershire 21
Brathay River 71
Breast-shot water wheel 28, 30, 37, 53, 77, 80, 87, 89–90, 90–3, 95, 97, 98, 98, 100, 101, 105, 107, 114, 117, 132–3, 136, 143, 152, 170
Brindley, James 87
Bristol 60, 67, 72, 86, 144, 155
Bristol Brass Co 67
British Aluminium Company 171, 172
British Electricity Authority 168
Bromford, Staffordshire 67
Bromsgrove 67
Bryanston estate 120
Buckets, water wheel 29–30, 30, 92, 93, 93, 102, 105
Buckfast Abbey, Devon 182
Buckinghamshire 67, 69, 128
Bude canal 87
Burghfield, Berkshire 134, 134
Burnhead dam 115

Buscot Park 94, 120
Buxton Mill, Norfolk 132
Byland Abbey 37
Byrncrug, Merionethshire 158

Calder River 79, 115, 139
Calne Water 60
Cam 45, 47, 47, 58, 65, 99
Cam River, Gloucestershire 60
Cambridgeshire 22, 23
Campbell, Robert 94, 120
Canals 15, 34, 43, 87, 89, 118, 119, 129, 131, 134, 144, 159, 165
Canterbury 19, 42, 129, 130
Carding 74, 77, 78, 79
Carr, John 124
Carron Ironworks 102
Carshalton, Surrey 102
Carus-Wilson, Eleanor 57, 58, 59, 60
Castle Donnington, Leicestershire 124
Castleford, Yorkshire 46
Catrine cotton mills 78, 91, 101, 106, 107, 108, 113, 142, 143, 179
Caudwell's Mill 120, 156, 156
Central Electricity Generating Board 168
Chalford, Gloucestershire 39, 79
Chard, Somerset 118, 152
Chart, Kent 19
Chatsworth, Derbyshire 35, 124, 126
Cheddleton, Staffordshire 85, 178
Chelmer River 129
Chelmer-Blackwater Navigation 129, 131, 134
Chelmsford, Essex 56, 134, 155
Cheshire 20, 46, 49, 50, 56, 64, 73, 76, 93, 143, 165, 166, 172, 182
Chesters, Northumberland 18
Chesterton, Warwickshire 111
Chiddingly, E. Sussex 131
Chingley, Kent 65
Chirk, Denbighshire 160
Churnet River 85, 160
Clasp-arm wheel 30, 31, 111
Clatteringshaws Lock 174
Claverton, Somerset 87
Clyde River 166, 173
Coalbrookdale 67, 83
Cold Edge Dam Co 115
Colne River 50

Colthrop mill 62
Company of Mines Royal 68
Compass-arm wheel 30, 95
Congleton, Cheshire 63, 63, 64, 74, 114, 166
Coniston 96
Constable, John 13, 14
Cookham, Berkshire 22
Cork 82
Corn mill 12, 13, 17, 20, 21, 22, 24, 27, 31, 32, 44, 45, 46, 47, 49, 56, 57, 58, 61, 62, 63, 67, 68, 69, 74, 78, 82, 84, 86, 88, 89, 100, 101, 106 ,108, 111, 112, 113, 116, 120, 124, 127, 128, 130, 131, 133, 138, 140, 144, 152, 155, 165, 169, 177, 178, 180
Cornwall 20, 32, 50, 60, 66, 68, 85, 87, 95, 96, 153
Cost of water power 27, 68, 114, 142, 143–5, 146, 149, 152, 182, 183, 184
Cost of water power – capital 12, 18, 22, 25, 42, 46, 78, 91, 101, 116, 130, 135, 143, 176
Cost of water power – running 22, 136, 137, 145, 151
Cost of water power – water wheel 46, 91, 97, 104
Costa Beck 161
Cotchett, Thomas 63, 72
Cotton industry 14, 64, 69, 74–8, 79, 80, 83, 113, 130, 138, 142, 144, 146, 147, 149, 166
Cotton mill 44, 73, 76, 77, 98, 100, 106, 107, 108, 112–3, 114, 140, 141, 142, 143, 165
Coulton 30
Council for the Protection of Rural Wales 168
Cragside, Northumberland 119, 169, 170
Cressbrook mill 114, 124
Cromford, Derbyshire 14, 39, 74, 75, 76, 82, 88, 103, 112, 114, 123, 139, 166, 180
Crompton, Samuel 76, 142
Crown wheel 111
Croxley, Hertfordshire 69, 151
Croydon, Surrey 159–60
Cutlery 67, 84, 152

Dale Abbey 41, 42
Dale, David 76

205

Index

Dam 34, 38, 40, 80, 84, 114, 115, 116, 136, 152, 167, 172, 174, 175
Danby, Yorkshire 129, 139
Daniel's Mill 129, 130
Darby, Abraham 67, 83, 102
Darby, H. C. 20
Darkley flax mill 96, 97
Dartford 66, 70, 154
Dartmouth, Earl of 79, 115
Darwin, Erasmus 167
Dean River, Angus, Scotland 43
Dean River, Cheshire 165
Deanston, Perthshire 78, 91, 101, 107, 108, 179
Dee mills 130, 172
Dee River, England 40, 42, 172
Dee River, Scotland 174
Defoe, Daniel 48, 55, 63, 67, 70
Derby 42, 63, 72, 73, 74, 75, 88
Derbyshire 11, 24, 36, 40, 41, 68, 73, 74, 76, 86, 88, 109, 114, 120, 124, 128, 139, 145, 156, 166, 169, 179, 182
Derwent mines, Co Durham 86
Derwent River, Derbyshire 35, 36, 41, 63, 69, 75, 76, 88, 90, 114, 139, 145, 167
Desaguliers, John 32, 88, 107
Devon Great Consols 153
Devon, River 43
Devonshire 20, 50, 56, 62, 66, 81, 84, 87, 94, 96, 153, 156, 169, 170, 178, 182, 184
Devonshire, Duke of 114
Dickinson, John 151
Dinorwic, North Wales 168
Disputes re mills 42, 43, 44, 52, 114, 139
Distilling 87
Dolgarrog 171, 172
Domesday Book 13, 19, 20–1, 22, 23, 26, 27, 40, 49, 50, 52, 65, 73, 131, 180
Dorset 20. 23, 46, 49, 118, 120, 128, 136, 137, 150, 155, 160
Down, Co 25
Downton, Wiltshire 21
Drainage, domestic and urban 24, 71, 159
Drainage, land 43, 160
Drainage, mines 68, 74
Drought 32, 53, 77, 80, 82, 138, 139, 140, 172
Droxford, Hampshire 159
Durham 23, 68, 86, 151
Dursley, Gloucestershire 60, 62

Eagley mills 113, 114, 116, 143, 146
East Anglia 55, 56, 57, 61, 62, 63, 90, 132, 133, 134, 144, 155
Ebley mill 44, 80
Ecclesbourne River 73
Edge tools 67, 83
Efficiency of wheels 18, 25, 27, 30, 31, 38, 74, 77, 86, 88–91, 93, 94, 95, 99, 100, 101, 102, 103, 117, 132, 133, 170
Egerton mill 96–7, 113, 146
Elan Valley, Mid-Wales 182

Electricity Board for Northern Ireland 170, 174
Eling tide mill 53, 53, 94, 113
Eliot, George 44
Elland, Yorkshire 44
Elsing, Norfolk 134
Elton, John 17, 177
Esgair Moel woollen mill 178
Esk Energy 183
Esk River, Yorkshire 41, 129, 139, 183
Evelyn, John 73, 165
Eversley, Hampshire 33
Ewart, Peter 77, 103
Ewelme River 60
Exe, River 42, 60, 136
Exeter 43, 56, 60, 69, 152, 165
Exmoor 86, 136, 137

Fairbairn, William 48, 78, 88, 90, 91, 83, 95, 96, 97, 100, 106–7, 108, 112, 113, 115, 138, 142, 146, 159
Fairbourne, Kent 56, 129
Fall [see also Head of water] 12, 25, 29, 31, 32, 34, 38, 54, 59, 68, 77, 78, 82, 88, 89, 90, 95, 98, 100, 101, 106, 107, 113, 114, 115, 120, 123, 132, 140, 143, 170, 171, 172, 183, 184
Falls of Clyde 173
Farey, John 24, 43, 90, 167
Farms and water power 12, 71, 118, 119, 120, 122, 134–7, 160, 169, 171, 181
Faversham, Kent 71
Feed milling 12, 111, 127, 137, 157–9
Fens 21
Fenton, Murray & Wood 143
Ffestiniog pumped storage 168, 175
Ffestiniog Railway 87, 168, 172
Finch, William 84
Finch, William Coles 177
Fishing and mills 42, 43, 45, 114, 157, 159, 160, 161, 167, 168, 183
Flash lock 43
Flatford mill 14
Flax 64, 70, 81, 82, 96, 118, 138, 145, 150, 171
Flint mills 85, 124, 178
Flintshire 113
Floating mill 33, 34, 55
Floats, water wheel 15, 30, 52, 91, 93, 94, 95, 97, 102, 183
Flood 32, 33, 44, 82, 91, 97, 138, 139, 140, 152, 158, 167
Flour milling 12, 13, 21–2, 23, 47, 54–6, 58, 73, 99, 111, 127–34, 138, 153–9, 179, 180
Forbes, R. J. 12
Ford 25
Forest of Dean 37, 65–6
Fort William 172, 184
Foulds, John Torr 72
Fountains Abbey 24, 24, 41, 49, 86
Fourdrinier machine 151
Fourneyron turbine 117, 117, 118
Frame, spinning 74, 76, 123, 142
Francis turbine 119

Frodsham, Cheshire 130
Frome River 60, 67
Frome, Somerset 84
Fulling 13, 22, 24, 31, 32, 44, 47, 47, 48, 52, 57, 58–63, 59–60, 64, 67, 68, 69, 71, 78, 79, 80, 81, 88, 140, 165
Funtley, Hampshire 150
Furness 67, 74, 113, 181
Fussell, James 84

Galloway 113, 174
Galloway water power company 173
Gardner, Samuel 118
Gearing, Gears 12, 16, 17, 25, 26, 44–7, 45, 48, 49, 52, 63, 72, 78, 88, 95, 99, 101, 102, 103, 104, 105, 106, 107, 108–12, 108–111, 113, 116, 126, 132, 133, 142, 155, 179
Giants Causeway 171
Gibson Mill, Yorkshire 147, 149
Gig mill 61, 80
Gilbert Gilkes & Gordon 119, 120, 171
Gilkes turbine 77
Gilkes, Gilbert 119
Glamorgan 65
Glamorganshire canal 87
Glasgow 115, 118
Glendoe, Highlands 168, 181, 181
Gloucester 41
Gloucestershire 20, 21, 33, 48, 59, 60, 80, 81, 114, 131, 142, 143, 144, 150, 170, 180
Glynn, Joseph 88, 98
Gnossal, Staffordshire 135, 136, 186
Godalming, Surrey 118, 170
Governor 82, 99, 116
Goyt River 182, 183
Grampian region 173
Grant River 170
Grassington, Yorkshire 86, 115, 167
Great Shelford, Cambridgeshire 22
'Greek mill' 17
Greenock 115, 170
Greg, R. H 114
Greg, Samuel 44, 76, 77, 147, 165, 166
Greg, Samuel junior 166
Grist mill 61, 158
Guildford, Surrey 55
Guisborough, Yorkshire 20, 111
Gunpowder 47, 52, 58, 59, 70–1, 70, 73, 114, 165, 167, 181
Gunton, Norfolk 86
Gypsum 13

Haden Brothers 143
Halton, Lancashire 89
Haltwhistle, Northumberland 18
Hammer mills 47, 58, 65–6, 84
Hammer pond 163
Hampshire 20, 21, 56, 64, 69, 94, 119, 131, 135, 152, 155, 156, 159
Hand mill 12, 13, 22, 23, 46
Handforth, Cheshire 93
Hargreaves, James 142

Hawick, Borders 148
Haworth, Yorkshire 79
Haxted, Surrey 132
Head of water [see also Fall] 12, 31–4, 37, 53, 91, 94, 95, 115, 116, 118, 121, 140, 183
Headrace 38, 78
Hebden Bridge, Yorkshire 149
Hele mill, Devon 55
Helmshore, Lancashire 60, 62, 105, 124
Hertfordshire 22, 69, 126, 152, 177, 183
Hessenford, Cornwall 158
Hett, Charles 120, 170
Hewes, Thomas 88, 99, 103, 106, 108
Highlands (Scotland) 167, 171, 174
Hollander 70
Holmes Reservoir Commission 118
Holy Brook 41, 161, 162–3
Holywell, Flintshire 143
Horizontal water wheel 12, 17, 18, 19, 25, 26–8, 26, 27, 33, 44, 45, 50, 116, 117, 177
Horse gear 68, 74, 119, 135
Horstead, Norfolk 2
Houghton, Huntingdonshire 21
Houston, Renfrewshire 93
Howsham, Yorkshire 43, 124
Huddersfield 62, 78, 145
Hull 70, 155, 156
Huntingdonshire 21
Huntley & Palmer 159
Hydraulic engine 86, 121, 122, 170
Hydraulic ram 121–3, 122, 137
Hydroelectricity 12, 13, 34, 107, 116, 119, 120, 167–8, 169–76, 169, 173–5, 180, 181, 182, 183, 184, 186

Ickham, Kent 18, 57
Industrial Revolution 30, 32, 38, 49, 54, 58, 63, 64, 73–87, 90, 95, 101, 102, 104, 112, 114, 115, 128, 133, 167, 186
'Industrial revolution', medieval 57–60
Invermay House 95
Ireland 17, 18, 19, 24, 26, 50, 51, 77, 81, 84, 94, 96, 100, 118, 150, 152, 155, 156, 158, 170, 176
Iron industry 30, 32, 38, 65, 74, 131, 151, 153, 163
Iron wheel 52, 72, 95, 95, 97, 101–6, 102–4, 108, 133
Irwell River 113, 114, 147
Itchen River 86, 164, 164

Johnston, Tom 173

Kames, Lord 22
Kaplan turbine 183
Keighley, Yorkshire 79, 80, 149
Kendal, Westmorland 71, 114, 116, 119, 137, 149
Kennet & Avon Canal 87, 131, 134
Kennet River 41, 58, 62, 128, 131, 133–4, 159, 161
Kent 14, 18, 19, 20, 50, 55, 56, 57,

Index

65, 66, 68, 69, 71, 127, 129, 152
Kent, River 71
Kentmere Head dam 116
Kinlochleven 171, 172, 175
Kinordy Lake 43
Kirkstall Abbey 65

Lady Isabella wheel 96, 105
Lancashire 20, 56, 62, 74, 75, 76, 80, 87, 89, 113, 114, 124, 130, 135, 142, 146, 147, 149, 152
Lantern wheel 46, 46, 105, 108, 109, 111
Launder 37, 38
Laxey, Isle of Man 96, 96
Layshaft drive 109, 109, 111, 112
Lea River 52, 71
Lead 68, 73, 74, 75, 86, 96, 115, 121, 154, 163, 165, 167, 171
Lealholm, Yorkshire 159
Leat 19, 29, 37–8, 41, 42, 43, 65, 66, 80, 114, 136, 142, 143, 145, 161, 164, 165, 167, 168, 181
Leeds 24, 65, 78, 80, 138, 140, 144, 145
Leen River 114
Len River 69, 129
Lenwade, Norfolk 134
Leven River 71
Lewis, M. J. T. 16, 25
Lillie, James 108, 112, 143
Limavady Electricity Supply Co 174
Lincolnshire 120, 127–8, 132, 136, 152
Linen 64, 81–2, 149
Liverpool 56, 154
Llantrisant, South Wales 65
Loddon River 21, 128, 159
Loft & Walne 136
Lombe, John 63
Lombe, Thomas 63, 64, 72
London Bridge 32, 34, 71–2, 72, 102
Lott & Walne 136
Loudwater, Buckinghamshire 62
Lough Island Reavy 115
Lumbutts mill 98, 99, 146
Luttrell psalter 30, 31, 48, 49
Lymm, Cheshire 76
Lynmouth, Devon 170, 180

MacAdam, Robert 118
Macclesfield, Cheshire 63, 64, 72, 74, 114, 145
Maentwrog power station 172, 175
Mainstream wheels 33, 41
Makeney, Derbyshire 41
Manchester 56, 73, 74, 76, 78, 82, 87, 106, 108, 141, 142, 143, 145, 146, 147, 165, 166
Manchester Statistical Society 147
Manorial mills 20, 22, 23, 24, 26, 54, 57, 62, 69, 84
Mapledurham, Oxfordshire 180, 180
Marlingford, Norfolk 134
Marlow, Buckinghamshire 55, 56, 67, 70
Marshall, John 138, 140, 145

Mary Tavy, Devon 96, 169, 170
Masson mill 36, 36, 41, 69, 75, 76, 124, 125, 139, 167, 180
Matlock, Derbyshire 75, 167
Medway River 55
Mersey Barrage Company 175
Mersey mills 130, 154
Mersey River 73, 82, 147, 175
Merton, Surrey 58, 165
Micro-hydro 181, 182
Middle Ages 13, 20–4, 26, 27, 28, 30, 46, 47, 57, 61, 62, 63, 65, 67, 73, 81
'Midget' roller mill 156, 157
Mill buildings 20, 22, 24, 25, 48–9, 62, 76, 80, 123–6, 161, 165, 179
Mill cottages 124, 165, 166
Miller 20, 22, 23, 24, 26, 33, 37, 43, 44, 45, 49, 53, 54, 55, 56, 57, 61, 85, 93, 104, 107, 109, 111, 115, 116, 127, 129, 130, 131, 132, 133, 134, 138, 153, 154, 155, 156, 157, 158, 159, 165
Millpond 12, 22, 34, 36, 37, 38, 41
Millstone 13, 16, 25, 26, 31, 44, 45, 46, 49, 108, 109–10, 111, 131, 136
Millwright 32, 44–5, 47–8, 72, 80, 88, 91, 95, 104, 105, 106, 107–8, 118, 119, 133
Milnthorpe, Cumberland 69
Mining 13, 32, 57, 58, 68, 74, 75, 83, 86, 90, 95, 96, 114, 120, 121, 139, 144, 145, 153, 163, 167, 168
Mole River 96
Monaghan, county 82
Monastic mills 23–4, 41, 50, 59
Montgomeryshire canal 87
Moorwhelham, Devon 87
Mule, spinning 76, 123, 142
Multure 22–3
Mumford, Lewis 12
Muncaster, Cumberland 57

Nailsworth, Gloucestershire 180
National Association of Water Power Users 160, 181
National grid 170, 172, 174, 181, 182
National Trust 178
Needham, Joseph 15
Needle-making 67, 67, 84, 152
Neme River 170
Nendrum, Ireland 18, 50
Nene River 21
Nether Alderley, Cheshire 46, 46, 49
New Lanark 76, 120, 124, 160
New Mills, Chseshire 35, 76, 183
Newark, Nottinghamshire 43
Newbridge, Sussex 65
Newbury, Berkshire 22, 54, 55, 56, 58, 60, 62, 69, 106, 128, 134, 155
Newcastle upon Tyne 72, 85, 127, 154, 170
Newcomen, Thomas 68, 73, 83, 89, 140
Newland Priory 38

Newland, Exmoor 86
Newlands, Keswick 68
Newnham, Northamptonshire 178
Nidderdale 148
Norfolk 20, 21, 86, 152
Noria 15, 16, 16, 71
'Norse mill' 17
North of Scotland Hydro-Electric Board 173, 175
North Wales Power & Traction Co 171–2
Northern Ireland 18, 50, 81, 87, 149, 155, 159, 170, 174, 175
Northumberland 23, 27, 130
Norwell, Nottinghamshire 23, 130
Nottingham 74, 106
Nottinghamshire 23, 68, 114, 130, 142, 143, 164
Nuneaton, Warwickshire 107, 108, 112

Old Windsor 18
Orkney 27, 185
Ouse Navigation 131
Ouse, Great, River 21
Overshot water wheel 16, 28, 29, 29, 30, 37, 46, 75, 83, 84, 86, 88, 89–90, 91–3, 96, 98, 100, 107, 108, 114, 117, 122, 136, 137, 140, 169, 171, 179, 183
Overton, Hampshire 64, 69
Oxford 60, 154, 155
Oxford Beck 161
Oxfordshire 182
Oyster wave energy converter 184–5

Painswick, Gloucestershire 62
Paper making 13, 27, 32, 58, 59, 61, 62, 67, 68–70, 69, 83, 88, 96, 116, 119, 120, 131, 139, 144, 151–2, 151, 165, 167, 171
Parent, Antoine 88–9
Paul, Lewis 74
Pelton wheel 119, 120, 121, 170, 171, 172
Pembrokeshire 40, 50, 53, 149
Pennines 62, 74, 76, 79, 80, 81, 82, 86, 115, 121, 140, 144, 145, 147, 149, 159, 163, 165, 167
Penstock 25, 38
Pentrough 37
Perth 91, 95, 98
Perthshire 91, 95, 98
Pewsey, Wiltshire 21
Philo of Byzantium 16
Pitchback wheel 37, 91, 92, 103, 130, 136
Poncelot wheel 53, 93–5, 94, 183
Poncelot, Jean Victor 94
Portal, Henri 69
Pound 34, 45, 53, 114
Pound lock 34, 45, 53, 114
Power of water wheels 32, 100
Power station 12, 34, 167–8, 170, 172–3, 174, 176, 180, 181, 182
Power, industrial 145ff
Preservation 96, 178–9
Price, Sir Uvedale 14, 167

Pumped storage 168, 170, 171, 175
Pumps 12, 15, 34, 45, 68, 71, 73, 75, 86, 87, 89, 94, 96, 108, 119, 120, 121, 123, 137, 138, 140, 141, 153, 178, 181

Quarme River 136
Quarry Bank mill 42, 44, 76, 77, 97, 100, 101, 103, 120, 124, 139, 142, 143, 145, 165, 178
Quarrying 71, 86, 171
Quern 12, 12, 21, 23, 25

Ramsbury, Wiltshire 161
Ramsden family 62
Reading, Berkshire 41, 56, 62, 128, 134, 155, 159, 161, 180
Redditch, Worcestershire 67, 67, 84, 152
Rees, Abraham 43, 90, 97
References in bold are to illustrations
Renewables obligation 182
Rennie, John 32, 85, 87, 92, 93, 106, 131, 134
Reservoirs 15, 34, 37, 44, 53, 71, 78, 80, 82, 83, 85, 89, 91, 98, 113–6, 121, 130, 136, 137, 139, 140, 142, 143, 146, 159, 160, 163, 167, 168, 170, 172, 174, 175, 182
Retford, Nottinghamshire 134
Reynolds, Terry 17, 32, 34
Rievaulx, Yorkshire 41, 65
Rights to water 42, 44, 64, 74, 75, 113–4, 127, 131, 139
Robertsbridge, E. Sussex 131
Robinson, George 114, 139, 142, 143, 144, 164
Rochdale, Lancashire 62, 87, 113
Roller milling 155–6, 158, 159, 177
Roman mills 12, 15, 17, 18, 32, 45, 46
Rossendale, Lancashire 62
Rossett Mill 50
Rotherhithe, Kent 50, 70
Royal Society 89
Ruswarp, Yorkshire 41, 41, 109, 183
Rynd 26, 46

Saddleworth, Lancashire 80, 145
Sagebien wheel 183
Salisbury, Wiltshire 60, 63, 106
Saltwater Bridge, Northern Ireland 52
Sarehole Mill, near Birmingham 84, 85, 110
Savery, Thomas 68, 73, 85, 100, 141, 140
Sawmill 49, 57, 58, 86, 86, 137, 170, 174, 182
Saxon mills 18, 20, 22, 26, 41, 50
Scotch turbine 118
Scotland 22, 24, 27, 43, 61, 68, 70, 78, 81, 86, 95, 108, 135, 143, 148, 149, 151, 168, 170, 172, 173, 175, 181
Scribbling 78–9
Scutching 64, 82
Scythe blades 67, 84

207

Index

Seed-crushing 70
Servern barrage 176, 184
Severn River 40, 81, 83, 175, 176, 184
Sexton, John 70
Sheaf River 67, 84
Sheffield 65, 67, 73, 74, 84, 86, 129, 152, 178
Shetland 26, 27, 56, 177, 184
Shipley, Yorkshire 79
Shropshire 65, 83
Shudehill, Manchester 76, 112, 141
Silk 63–4, 63–4, 72, 73, 74, 82, 114, 124, 150, 151, 166, 182
Silk reel 111
Slate quarries 71, 171
Sliding hatch 92–3, 99
Sluice gate 25, 38, 42, 92, 92, 93, 99, 160
Smart, Maurice 58
Smeaton, John 27, 30, 32, 44, 48, 72, 88–93, 95, 99, 102, 106, 111, 117, 121, 141
Snuff 58, 86, 165, 167
Society for the Protection of Ancient Buildings 31, 177
Society of Mineral and Battery Works 67
Soke mills 22, 24
Somerset 65, 70, 81, 96, 137, 150, 152, 170
Sorocold, George 48, 63, 72
South Wingfield, Derbyshire 169
Sowerby Bridge, Yorkshire 60, 148
Spade mills 67, 84, 152, 167
Spilman, John 70
Spinning frame 74, 75, 76, 123, 142
Spinning jenny 142
Spodden River 113
Spurwheel 111–2, 110, 133
Staffordshire 19, 20, 21, 65, 67, 70, 85, 136, 140, 160
Standon, Hertfordshire 22
Stanley mills 80, 125, 142
Staveley, Westmorland 87
Steam engines 13
Steam power 13, 53, 68, 69, 71, 73–4, 77, 78, 79, 80, 81, 82, 83, 85, 86, 87, 89, 90, 94, 99, 100, 107, 113, 114, 115, 116, 117, 127, 131, 136, 137, 128–9, 170, 172, 176
Steam pump 141
Sticklepath, Devon 83, 84, 178, 179
Stockport 63, 74, 82, 113, 114, 145
Stockport Hydro Ltd 182
Stone cutting 57, 86
Stour, River (Suffolk) 14, 14, 56, 107, 132, 134
Strabo 16, 18
Strangford Lough 50, 52, 175
Stroud, Gloucestershire 44, 60, 80

Strutt, family, cotton millers 14, 74, 82, 88, 97, 166, 181
Strutt, Jedediah 74, 76, 82, 106, 114, 116, 139
Sturminster Newton, Dorset 23, 128
Styal, Cheshire 44, 76, 101, 108, 114, 120, 142, 145, 166, 179
Suffolk 14, 31, 52, 53, 107, 127, 132, 144, 178
Surrey 58, 71, 73, 102, 118, 129, 151, 170
Suspension wheel 95, 104, 105–6, 108, 143
Sussex 53, 56, 131
Sutton Courtney, Berkshire/Oxfordshire 69
Sutton Poyntz, Dorset 118
Sutton under Whitestonecliff, Yorkshire 20

Tadmarton, Oxfordshire 19
Tailrace 32, 37, 44, 78, 83, 100, 101, 115, 180
Tamar River 87
Tame, River 21, 31, 67, 70, 131
Tamworth, Staffordshire 19
Tan-y-Grisiau, Mid-Wales 168
Tate, John 68
Telford, Thomas 83
Temple Newsam, Wiltshire 59
Test River 64
Thames River 18, 34, 40, 41, 43, 50, 55, 56, 67, 70, 71, 72, 128, 133, 151, 159, 180, 182
Thirlage 22, 24
Thomson, James 118
Thorrington, Essex 50, 51, 52, 53
Three Mills, Bromley by Bow 50, 52, 52, 94
Threshing machine 135–6, 137
Tidal current generator 184
Tidal power 175–6, 184
Tide mills 18, 49–53, 51–3, 94
Tillingbourne River 58, 71, 73, 165
Tillingham River 131
Tin 32, 66, 68, 96
Tintern, Monmouthshire 67
Tolls, milling 22–3, 54, 55
Torridge, River 156
Torrington, Devon 62
Torrs Hydro 182, 183, 183
Traditional Corn Millers Guild 180
Transport 55, 56, 79, 86, 87, 104, 105, 129, 133, 134, 144, 154, 155, 161, 172
Trent River 40, 42, 43, 85, 124
Trip hammer 45, 65, 66, 69
Trowbridge, Wiltshire 81, 143
Trundle gear 46, 108
Tummel power station 173, 175
Turbine 13, 27, 49, 71, 77, 82, 87, 93, 94, 95, 100, 107, 108, 116–21, 116–21, 123, 137, 140, 146, 148, 150, 152, 156, 157, 166, 168, 169, 170, 171,
172, 174, 180, 181, 182, 183, 184
Tutbury, Staffordshire 101, 140
Twyford, Berks 64
Twyford, Hants 135

Ulster 31, 65, 77, 81, 82, 115, 118, 150, 155, 171
Undershot water wheel 15, 26, 28–30, 29, 32, 33, 38, 44, 53, 63, 65, 80, 88–92, 93, 94, 95, 98, 100, 129, 132, 141, 150, 183
Usk, Gwent 160
Uxbridge, Middlesex 165

Vitruvius 15, 16, 17, 30, 45
Vortex turbine 77, 118, 119, 118–19, 171

Wakefield mill 102
Wakefield, John 71
Wakefield, Yorkshire 24, 89
Wales 24, 32, 60, 70, 74, 81, 87, 98, 103, 143, 148, 149, 168, 170, 172, 175, 180, 182
Wallingford, Oxfordshire 56
Waltham Abbey, Essex 71, 98
Wandle River 44, 58, 64, 129, 159, 160, 165
Wandsworth, Surrey 64, 165
Wansford, Northamptonshire 69
Warminster, Wiltshire 21
Warrington, Cheshire 56, 130, 154
Warwick 129
Warwickshire 21, 108, 111
Washburn River 145
Water Resources Act 1963 160, 181
Water supply, for mills 21, 30, 32, 33, 42, 44, 71, 74, 80, 82, 84, 112–16, 130, 131, 135, 138–40, 144, 146, 160, 162
Water supply, public 71–2, 159–60, 161, 182
Watermills, numbers 19, 20–1, 58, 59, 60, 62, 113, 155, 156
Waterwheel 25–32, 28, 31, 88–107, 106, 136, 154
Waterwheel construction 101 ff
Waterwheel size 31–2, 95–7
Watt, James 73, 85, 99, 100, 121, 127, 140, 141–3
Wave energy 185
Waveney River 56, 134
Weald 19, 65
Wealden iron industry 38, 65–6, 131, 163
Weaving, cotton 78, 142
Weaving, linen 82
Weaving, silk 63
Weaving, wool 80, 142
Wedgwood, Josiah 85
Weir 11, 11, 19, 34–43, 35–6, 41–2, 77, 78, 91, 114, 115, 130, 140, 143, 160, 161, 164, 167, 172, 181, 182, 183

Welsh Water 160
Welshpool, Powys 87
Wensum River 134, 152
West Country 24, 48, 56, 58, 61, 79, 80, 81, 84, 91, 143, 148, 149, 150, 152
West Devon Electric Supply Company 170
West Harnham, Wiltshire 61
Westmorland 71, 73, 87, 148
Weston-upon-Trent, Derbyshire 13
Wey River 55, 69, 170
Weybridge, Surrey 151
Weymouth Waterworks Co 118
Wharfe River 79
Wharram Percy 18, 22
Wheel houses 123, 123
Whitchurch, Hampshire 64, 64, 69, 124
Whitelaw, James 118
Whitmore & Binyon 130, 133
Wickham, Hampshire 159
Wickwar, Gloucestershire 33, 170
Williams-Ellis, Clough 168
Williamson Brothers, Kendal 119, 137, 170
Wilton, Wiltshire 63, 131
Wiltshire 20, 21, 33, 56, 60, 81, 86, 131, 144
Wimborne, Dorset 160
Winchester, Wiltshire 46, 64, 67, 135, 164
Windmills 21, 22, 25, 41, 53, 54, 64, 70, 89, 99, 102, 111, 127–8, 131, 154, 158, 177, 180
Wirksworth, Derbyshire 76, 114, 139, 141
Woodbridge, Suffolk 53, 179
Woolhanger, Devon 137
Woollen industry 57, 59–60, 61, 62, 64, 70, 74, 78–81, 83, 142, 144, 147–9, 150
Woollen mills 44, 80, 84, 115, 130–1, 139, 144, 148, 150, 152, 178
Worcestershire 65, 67, 84
Worsted 63, 74, 78, 79–81, 115, 144, 147, 148, 149
Wotton, Surrey 71, 73
Wright, Joseph, of Derby 14
Wye River, Buckinghamshire 69
Wye River, Derbyshire 44, 114, 124
Wylie River 21

Yale Electric Power Co 171
Yare River 134
Yorkshire 18, 29, 30, 37, 38, 41, 43, 44, 46, 49, 55, 59, 60, 68, 70, 74, 80, 81, 86, 87, 102, 109, 111, 112, 114, 115, 124, 129, 139, 144, 145, 147, 148, 149, 159, 161, 182, 183
Ystrad River 172

Zuppinger wheel 183